COMMUNICATIONS IN THE TWENTY-FIRST CENTURY

This book is based on a symposium sponsored by The Colgate Darden Graduate School of Business Administration at the University of Virginia in cooperation with The Annenberg School of Communications at the University of Pennsylvania and The Annenberg School of Communications at the University of Southern California.
It is the second in a series of Twenty-First-Century programs funded by Philip Morris Incorporated. The first title in the series is Working in the Twenty-First Century.

COMMUNICATIONS
IN THE
TWENTY-FIRST
CENTURY

Edited by

ROBERT W. HAIGH
The Colgate Darden Graduate School of Business Administration
University of Virginia

GEORGE GERBNER
The Annenberg School of Communications
University of Pennsylvania

RICHARD B. BYRNE
The Annenberg School of Communications
University of Southern California

A WILEY-INTERSCIENCE PUBLICATION

JOHN WILEY & SONS

New York   Chichester   Brisbane   Toronto   Singapore

Published by John Wiley & Sons, Inc.
All rights reserved. Published simultaneously in Canada.

This publication is designed to provide accurate and
authoritative information in regard to the subject
matter covered. It is sold with the understanding that
the publisher is not engaged in rendering legal, accounting,
or other professional service. If legal advice or other
expert assistance is required, the services of a competent
professional person should be sought. *From a Declaration
of Principles jointly adopted by a Committee of the
American Bar Association and a Committee of Publishers.*

Library of Congress Cataloging in Publication Data:

Main entry under title:

Communications in the twenty-first century.

    Based on a symposium held in the spring of
1981 in Richmond, Va. and sponsored by The Colgate
Darden Graduate School of Business Administration
at the University of Virginia in cooperation with
The Annenberg Schools of Communications at the
Universities of Pennsylvania and Southern California.
    Bibliography: p.
    Includes index.
    1. Telecommunication—Congresses. 2. Communica-
tion—Congresses. I. Haigh, Robert William,
1926-   . II. Gerbner, George. III. Byrne,
Richard B. IV. Colgate Darden Graduate School of
Business Administration. V. Annenberg School of
Communications (University of Pennsylvania)
VI. Annenberg School of Communications (University
of Southern California)
HE7631.C57   380.3   81-14797
ISBN 0-471-09910-4   AACR2

Printed in the United States of America

10 9 8 7 6 5 4 3 2

CONTRIBUTORS

ELIE ABEL, LL.D., Harry and Norman Chandler Professor of Communication, Stanford University

GAIL E. BERGSVEN, Vice President, Human Services Programs, Control Data Corporation

ANNE W. BRANSCOMB, J.D., Communications Consultant

ASA BRIGGS, Lord Briggs of Lewes, LL.D., Litt.D., D.Sc., Provost of Worcester College, Oxford, and Chancellor of the Open University

HARRY L. FREEMAN, Senior Vice President, Office of the Chairman, American Express Company

LAWRENCE HALPRIN, Landscape Architect Planner; Founder, Round House

ELIHU KATZ, Ph.D., Professor of Sociology and Communication, The Communications Institute, Hebrew University of Jerusalem, and The Annenberg School of Communications, University of Southern California

PETER G.W. KEEN, D.B.A., Associate Professor, Management Science, Sloan School of Management, Massachusetts Institute of Technology

AMORY B. LOVINS, D.Sc., British Representative, Friends of the Earth; Consultant Physicist; Author

L. HUNTER LOVINS, J.D., Sociologist and Political Scientist

LOUIS H. MERTES, Vice President and General Manager, Systems, Continental Bank

J. RICHARD MUNRO, President and Chief Executive Officer, Time Inc.

KATHLEEN NOLAN, Actress; former President, Screen Actors Guild

KAARLE NORDENSTRENG, Ph.D., Professor of Journalism and Mass Communication, University of Tampere, Finland

ARNO A. PENZIAS, Ph.D., Nobel Laureate; Executive Director, Research, Communications Sciences Division, Bell Laboratories

JOHN P. ROBINSON, Ph.D., Professor of Sociology; Director, Survey Research Center, University of Maryland

F.G. RODGERS, IBM Vice President, Marketing, International Business Machines Corporation

v

HERBERT I. SCHILLER, Ph.D., Professor of Communications, Third College, University of California, San Diego

KIMIO UNO, Ph.D., Associate Professor, Institute of Socio-Economic Planning, University of Tsukuba, Japan

TIMOTHY E. WIRTH, U.S. Representative (D. Colorado)

FOREWORD

The exploration of the twenty-first century by Philip Morris Incorporated in the distinguished company of leading academic institutions is a most rewarding experience for our corporation. We have been afforded a rare opportunity to step aside for a moment from our daily preoccupations and, under the guidance of highly informed mentors, take a perceptive look at the momentous developments in which the corporation is involved with all of society.

The first in this series of Twenty-First-Century programs was held in the spring of 1979 at the Philip Morris Manufacturing Center in Richmond, Virginia. It was entitled "Working in the Twenty-First Century" and was cosponsored by The Colgate Darden Graduate School of Business Administration at the University of Virginia and The Wharton School at the University of Pennsylvania. The papers delivered at that conference were published in book form.

The second two-day symposium on the future, which has resulted in the present book, was held in the spring of 1981 in Richmond and was entitled "Communications in the Twenty-First Century." Once again The Colgate Darden School sponsored the event, this time in cooperation with The Annenberg Schools of Communications at the University of Pennsylvania and the University of Southern California.

The diversity of the speakers and the lively interaction that resulted were precisely what the planners of the symposium intended, and in this they were enthusiastically seconded by our company. The agreed-on purpose was not to find solutions or even establish a consensus but to attempt to define issues as a first step toward devising communications strategies that will work toward the common good for us and for those who follow.

It is our hope that we have helped to take this small but very important step toward the year 2001. All of us, and most particularly those of us in business, must be prepared.

George Weissman
Chairman of the Board
Philip Morris Incorporated

PREFACE

Communication is as old as humanity itself and at critical points in our history has transformed our lives. We live in such a time. The potential of the new communications technology is far reaching, but the important issues relate to how we can integrate these advances into our personal, business, and political lives.

We have learned in this century that technological progress does not necessarily mean human progress. The tools of communication have as much power to alienate people as to bring them together. This troubling duality is a characteristic of advances in the arts and sciences that we often choose to overlook in our scramble to make progress.

As the reader will discover, there are sharp disagreements among the contributors to this book over several of the issues. Such debate characterized the symposium at which they presented their views. It was planned that way. The conference, "Communications in the Twenty-First Century," sought to examine the future of communications in its broadest sense. The participants focused on such concerns as public policy, implications for the individual, strategies for management, the future of the media, and the high stakes involved in the free flow of information across international boundaries.

We were fully aware that there are no easy answers, but we are also convinced that only in free and frank exploration of the issues will we begin to find our way to their resolution. In his welcoming remarks at the conference, Ross R. Millhiser, Vice Chairman of the Board of Philip Morris Incorporated, summed up the compelling imperative:

> It is not just the technology of future communications that commands our attention but also the willingness of people and nations to use communications

to enlarge understanding. Information must be marshaled continuously for the advancement of society.

This book deals with some of the oldest and some of the newest themes of civilization. Julian Huxley, grandson of the great defender of Darwin's theory of evolution and a distinguished scientist in his own right, once noted that the invention of the phonetic alphabet more than 3000 years ago was one of those irreversible achievements that permanently alter the course of humankind. He also observed something else: No sooner was there a breakthrough to this great and simplifying principle than the tribes and peoples seized on it to multiply Babel into countless written languages and dialects. "The pursuit of the alphabet up and down the corridors of history," observed Huxley, "illustrates what seems to be a characteristic of human life—namely, the inseparability of the twin tendencies towards differentiation and towards integration."

Today we face another irreversible breakthrough made possible by what one of the contributors to this book calls "the constellation of inventions that converges in the marriage of computers and telecommunications." This remarkable development sometimes goes under the name of "compunications,"* a term used by several authors in the pages that follow to fill the gap in nomenclature. By any name this concatenation of technologies has produced what still another contributor calls "synthesis and individuation" on a scale never before imagined. Each author in the book, beginning with Elie Abel in the prologue, attempts in one way or another to come to terms with the implications for individuals and for society of the powerful polarity described by Huxley.

Associated with this is another question that seriously troubles Abel and other authors. It was first raised in succinct form by Abraham Lincoln when, in reaffirming his belief in the native wisdom of the people, he made his famous observation that, "If given the truth, they can be depended upon to meet any national crisis. The great point is to bring them the facts."

The new era of information presents an unparalleled opportunity to provide people universally with precisely those hard facts required for the survival of democratic institutions and processes. But, as Abel points out, there is reason to doubt whether the wide dissemination of knowledge and information implied will actually occur, despite the proliferation of the me-

*The word was coined by Anthony G. Oettinger, Chairman of the Program on Information Resources Policy at Harvard University.

dia, programs, and systems. Will vital information be distributed equally to all? Will much of what is actually communicated turn out not to be the kind of "facts" that Lincoln had in mind but mere babble not worth hearing?

Still another major theme weaves its way through the book, reappearing in different forms and contexts as it has in western society since it was bequeathed to us by the ancient Greeks: "Knowledge itself is power." Disraeli expressed the more modern view with charming candor when he said, "As a general rule the most successful man in life is he who has the best information." So with groups, institutions, and whole societies.

Information is of two kinds, says one of the authors: There is the kind of information that "diminishes uncertainty," and anything that does not fit into this category—a very great deal indeed—classifies as "all other." This is an economist's approach to the problem and in essence it satisfies the needs of bankers, technologists, engineers, the military, statisticians, government bureau chiefs, and other people who must deal with clear numbers and hard facts. Whoever controls the gathering, storing, and dissemination of this information commands a vitally important resource, one that will always be in scarce supply.

The implications are hardly lost on business competitors or on the nations of the world, particularly those of the Third World. Thanks to the power of modern information networks, a minor functionary today could conceivably have more information at his or her fingertips than Disraeli could have pulled together from a worldwide empire. Of course, the question is whether he would recognize the worth of that information or use it as effectively as Disraeli did when he cannily used a few key facts to diddle the French out of the Suez Canal. There is also the point made in this book that an information system is no better than the information that is put into it in the first place.

And what about that "all other" category, the part of communication that involves not just our intellects but the entire range of senses and sensibilities? What happens to the *whole* human being in a world in which electronic communications dominates our work as well as our leisure? This question is explored by several contributors who do not see communications only as a transfer of economic values or as a means of strategic control of resources but rather as an expression of our essential humanity. This, in turn, raises the issue whether in the long run humans will have to accommodate themselves to machines or whether the machines will be made to serve wholly human purposes.

And what of the future? In the epilogue to the book Asa Briggs agrees with Elie Abel in the prologue that the next ten or fifteen years will be critical in terms of decisions that can be expected to alter totally the course of communications. But Briggs goes even farther when he says:

> During the first two or three decades of the new century, the people making some of the key decisions will be those who have already finished their formal education. These people therefore will belong to this century as well as to the next; they will be the bridge generation.

> In that sense, the future is already here, as it is in other ways. Some trends, notably demographic, seem already to have been set. Certain managerial and political decisions concerning the twenty-first century, given the time lag in implementation, have already been taken. New generations will follow, however, that never knew what it was like to live in the twentieth century; they will be looking forward with mixed emotions to the twenty-second century.

We believe that readers who keep in mind the arresting thought that "the future is already here" will be doubly rewarded in reading the provocative writings in this book.

Robert W. Haigh, D.C.S.
George Gerbner, Ph.D.
Richard B. Byrne, Ph.D.

Charlottesville, VA
Philadelphia, PA
Los Angeles, CA
October 1981

ACKNOWLEDGMENTS

All books benefit from the care and contributions of many people. This volume is no exception. We acknowledge with thanks and appreciation the distinguished speakers who shared their ideas at the "Communications in the Twenty-First Century" symposium and on whose papers this book is based.

We also wish to thank our academic colleagues who worked so creatively in preparing an interesting and informative program: William E. Zierden, Ph.D., Associate Professor of Business Administration, from The Colgate Darden Graduate School of Business Administration; Barry Cole, Ph.D., Adjunct Professor of Communications, and Elvira Lankford, Assistant Dean, from The Annenberg School of Communications (University of Pennsylvania); and Professor Herbert S. Dordick from The Annenberg School of Communications (University of Southern California).

To Philip Morris Incorporated we extend thanks for funding the symposium and for the use of their splendid auditorium and other facilities in Richmond.

Special gratitude must go to that company's perceptive, patient, and helpful executives on the Conference Committee: George Weissman, Chairman of the Board; Ross R. Millhiser, Vice Chairman of the Board; Clifford H. Goldsmith, President; Hugh Cullman, Group Executive Vice President and Chairman and Chief Executive Officer, Philip Morris U.S.A.; James C. Bowling, Senior Vice President, Assistant to the Chairman of the Board, and Director of Corporate Affairs; W. Wallace McDowell, Vice President and Executive Vice President, Operations, Philip Morris U.S.A.; James A. Remington, Senior Vice President, Manufacturing, Philip Morris U.S.A.; Frank A. Saunders, Staff Vice President, Corporate Relations and Communications;

Vincent R. Clephas, Director, Corporate Public Affairs; James M. Frye, Director, Community Relations, Philip Morris U.S.A.; and Paul A. Eichorn, Ph.D., Conference Operations Coordinator and Assistant Director of Laboratory Administration, Philip Morris U.S.A. Joan Mebane, Philip Morris Incorporated Manager of Communications Research and the Conference Coordinator, organized the conference and guided the editorial production of this book with able assistance from Gina Gallovich, Communications Research Coordinator. Susan von Hoffmann, Communications Research Coordinator, provided editorial support.

Finally, our thanks to Carl Rieser, a former editor of *Fortune, Business Week,* and the Committee for Economic Development, whose editorial skills were deftly applied in working with our authors. His was a graceful contribution to the style and pace of the manuscript.

R.W.H.
G.G.
R.B.B.

CONTENTS

COMMUNICATIONS IN THE TWENTY-FIRST CENTURY

21

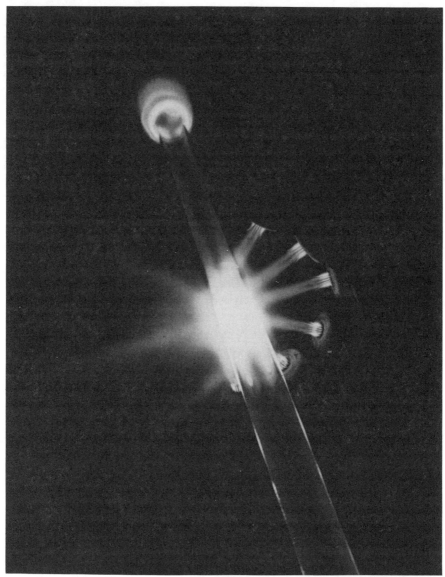

Optical fibers improve speed, capacity, and efficiency by transmitting information as bursts of light through glass fibers rather than as electrical signals through thicker copper wire cables. Courtesy of Bell Laboratories.

PROLOGUE

LOOKING AHEAD FROM THE TWENTIETH CENTURY

ELIE ABEL

The hazards of prophecy are incalculable. Consider, as we approach 1984, the case of George Orwell. For my generation 1984 was not so much a date as a metaphor, a nightmare vision of the ultimate totalitarian future. My students today see 1984 simply as a date, nothing more, marking the year in which many of them will take their undergraduate degrees and go out into what they call "the real world." That world of the 1980s, for all its frustrations and disappointments, happily falls short of the horrors described so chillingly by Orwell. If an appreciable number of present-day students read Orwell, which I am inclined to doubt, they may well conclude that Arthur C. Clarke had a firmer grip on the shape of the future.

Consider next the case of Goethe, who in one of his less inspired poems forecast a uniquely serene future for America. "Amerika, du hast es besser," Goethe wrote about 1820. In English, courtesy of John Lukacs, the poem goes like this:

> America, you're better off
> Than our ancient continent.

1

You have no dilapidated castles
No old stone for monuments.
Your soul, your inner life
Stays undisturbed by
Useless memory
And unprofitable strife.

Goethe had never seen America. He was looking ahead from the nineteenth century, a high-risk exercise even for that universal genius of Weimar. Because, of course, he got it wrong. Working from an inadequate data base, he foresaw an America untouched by the main currents of world affairs, a nation outside history, so to speak, that would be spared the "unprofitable strife" that was Europe's tragic lot. There was no room in Goethe's Arcadian vision of America for the Civil War, which broke out less than thirty years after his death, nor did he foresee the emergence of the United States as a world power, a nation no longer pure of soul.

I suspect that Philip Johnson was right when he suggested, on receiving the gold medal of the American Institute of Architects, that the shape of the future may be far less "modern" than our garden-variety missionaries keep telling us. Our sensibilities have changed in three important ways, Johnson said. He put it this way:

"Modern" hated history; we love it.
"Modern" hated symbols; we love them.
"Modern" built the same look in any location; we search out the spirit of the place—the *genius loci*—for inspiration and variation.

Johnson was speaking as architect to architect, but he ventured the heretical notion that progress and reason and the good life may not, after all, be the ultimate guideposts of American destiny. "We are entering an era," he said, "that I don't know the name of and even those that say they know the name of don't know the name of. But it's a great, adventurous, pluralistic future." There you have it: great, adventurous, and, above all, pluralistic and nameless.

Now that I have hedged my bets by reminding you how far off the mark were the prophecies of our betters, Goethe and Orwell among them, not to mention Marshall McLuhan, let us turn to the outlook for communications in the twenty-first century. Wiser men than I have made the observation that flows of information have accelerated at a fantastic rate since mankind started scratching records on clay tablets or on the walls of a sheltering cave. Wilbur Schramm, for example, has worked out a progression that looks like this:

From spoken language to writing: at least fifty million years.

From writing to printing: about 5000 years.

From printing to the development of sight-sound media (photography, the telephone, sound recording, radio, television): about 500 years.

From the first of the sight-sound media to the modern computer: fewer than fifty years.

What the next two decades will bring in the field of human communication not even Wilbur Schramm dares to predict. The year 2001 will be on us before most contemporaries will have digested the implications of the discoveries and inventions already accomplished. One lesson we *have* learned: that new technologies do not necessarily displace the older ones. Television did not kill radio or motion pictures, radio did not kill the gramophone, and the printed word is with us still. I do not mean to suggest that the advent of a new technology leaves the existing communications technologies wholly unaffected; only that the mix becomes richer, more variegated, and more versatile, one medium influencing another, as radio clearly influenced the shaping of television without the total sacrifice of its own characteristic identity.

What we have seen in the United States over the last three or four decades is a reallocation of functions among media that exist side by side, even as they compete for audience attention and advertising support. If the broad-spectrum national magazine has fallen by the wayside, there has been a flowering of special-interest magazines. If radio no longer occupies the center of the family circle of an evening, as it did when many of us were children, it travels with us to and from our daily work, keeping us, if not informed, at least alert to news developments, the weather, and road conditions. These intrusions of rude reality are interspersed with recorded popular or classical music, according to the listener's personal taste. Indeed, the success of all-news radio has already prompted a nationwide all-news television service for distribution by satellite to cable systems around the country.

The text revolution is on us today. It delivers on my doorstep in California at 7 A.M. a newspaper that was edited on a computer in New York and relayed by satellite from Massachusetts to a printing plant two miles up the road from my house in Palo Alto. The final stage of that cosmic journey is completed, rather prosaically, by a small boy on a bicycle, as it was in my grandfather's time. But that, too, will change before long, we are assured. News will be delivered to the home electronically on the family television screen or, for people who insist on their God-given right to mull at leisure over columns of words and numbers, on a black-on-white printout.

I do not derive much comfort or pleasure from this prospect. My attitude toward futurism is decidedly skeptical, perhaps borrowing something from Thoreau, who in 1854 wrote: "We are in great haste to construct a magnetic telegraph from Maine to Texas, but Maine and Texas, it may be, have nothing important to communicate." Many of Thoreau's contemporaries wrote him off as a crank. They believed that tomorrow held eternal promise, that the forward march of education, industry, agriculture, and transportation, along with quicker and more reliable communication, necessarily spelled human progress. In our time we have come to question the inevitability of man's progression toward the good life through chemistry, as one of our national advertisers used to assure the public day by day. Or, for that matter, through electronics, genetic engineering, or other scientific frontiers. The media are powerful and intrusive, endowed by the communications satellite with the power to annihilate distance. What about the message? What *does* Maine say to Texas or Paris or Ouagadougou?

Thoreau, if I understand him, would not have been content to leave the framing of that message to engineers and scientists. Nor indeed to the interplay of vast corporate interests, preoccupied as they are with their central function of producing goods or services at a profit. Least of all would he have trusted government to speak for him. Perhaps his most important essay, "Civil Disobedience," grew out of his experience in defying the U.S. government. You will recall that he spent one night in prison for refusing to pay a poll tax in support of the Mexican War, which he regarded as an effort to extend slavery. Thoreau held that civil disobedience was the individual's just means of protesting unjust actions by the state. And without the help of computers, satellites, or even radio waves his printed message reached around the world. It inspired Gandhi's passive resistance movement in India and more recently the nonviolent civil rights campaign led by Martin Luther King here at home.

If Thoreau could look in on our society today, 119 years after his death, he would certainly find the speed and reach of communications systems, as we know them, bewildering. He would, I am certain, be distressed by the essential sameness of the product whose increasing homogeneity in content as well as format is so upsetting to many social critics, and he would be even more dismayed by the trivialities we choose to disseminate. Our newspapers, no matter where published, get the bulk of their information from one of two large agencies. Our three commercial television networks use what appears to be a uniform standard of news selection, and a uniform standard of tastelessness is applied to what passes for popular entertainment. Our regional speech patterns and other traditional distinctions are fast disappearing as they are mashed and macerated in an enormous cultural blender. It would be difficult to find a corner of the country in which the basic diet of information about the state of the nation or the world differs significantly from the diet supplied a thousand miles away.

This is only partially the result of current technology. It is no less the product of the spread of chain ownership; literally dozens of newspapers that once were locally owned have now been taken over by large media corporations. These enterprises capitalize on certain economies of scale by collective purchasing, not only of newsprint and equipment but also, with rare exceptions, of information.

Not a reassuring picture, to be sure, and one that poses a question: What help can we expect from the new technologies as they take hold more widely? The answer is even less reassuring. It is likely that there will be further audience fragmentation and less common sharing of knowledge by rich and poor, by advantaged and disadvantaged. This could seriously erode the common data base that makes our system of self-government possible.

The daily news, as we know it, may be superficial and mostly bland, but it does provide the American people with a common pool of information, which is not to say that it produces uniform opinions or attitudes. The effect of the media's handling of the news is to thrust into the foreground many of the concerns that Americans will share sooner or later by telling them what to think *about*, if not what to *think*. If present trends are projected through the end of the century, we shall be living with rather different media. The bill of fare will be more varied and will permit wider freedom of choice. It will probably be more appropriate by the year 2001 to speak of narrowcasting rather than broadcasting, as individuals pick and choose those offerings that particularly appeal to them. This could mean, for example, a steady diet of sports or financial news or first-run motion pictures—all at a price—instead of the mélange now programmed by the TV networks.

Radio has already reached that stage: Listeners can choose among, say, country and western and hard rock music stations or between all-news formats and talk shows. Magazines reached that point of audience segmentation by the early 1970s. Trout fishermen, antique automobile fanciers, investors, liberated and unliberated women, skiers, golfers, stamp collectors, and tennis players all have their highly specialized magazines. What else these people may read is something of a mystery, but they no longer have the choice of subscribing to general-circulation magazines like *Collier's* or *Look*. One can, it is true, find *Life* and the *Saturday Evening Post* back on some newsstands today, but they are not what they were.

Back in my student days it was possible to identify a tiny pocket of mid-Manhattan, no more than two square miles in area, as the seat of the mighty—the place where all big media decisions were made. What Americans read and heard on the radio in those days was decided there, between 40th and 57th Streets, a few blocks each way to the east and west of Fifth Avenue. The National Broadcasting Company, Time Inc., and the Associated Press were in Rockefeller Center. The Columbia Broadcasting System was four blocks away at 485 Madison Avenue. The headquarters of *Look* and *Collier's* were nearby. *The New York Times* and *Herald Tribune* were in the West Forties and the *Daily News* and *Daily Mirror* were in the East Forties, along with the United Press Association, as it was called then. Many of the survivors among these giants of the communications world are still in the neighborhood, though not all at their old addresses. The biggest change, however, is the emergence of competing media power centers a long way from New York: Knight-Ridder in Miami; Gannett in Rochester, New York; Times-Mirror in Los Angeles; the Cable News Network in Atlanta.

Walking along the Avenue of the Americas you can feel the tension from the headquarters of NBC at 50th Street to CBS at 52nd and ABC at the corner of 54th. The men in charge have seen the future—and they fear it may work, chipping away at their aggregated lion's share of the national viewing audience.

What the networks have to sell is, of course, audience mass to advertisers. If their total audience were to shrink appreciably because of the competition of cable television and other newer technologies, significant losses of revenue and power would occur. Most network executives, when they speak for the record, still make light of the threat. "Toys for the wealthy," says a senior CBS executive. "Television as we know it . . . will be with us for a long time to come." But there are disinterested students of the industry who forecast a definite fragmentation of the national audience. Specifically, they predict that by 1990 network television will have lost no less than 10 percent and as much as 50 percent of its audience. For an industry that has known more ups than downs over the decades the effect is likely to be cataclysmic.

Over the next decade or two the sharp line that separates print from broadcast can be expected to blur. We can then expect to see the birth of hybrid multimedia information systems for the home that depend less on the support of advertisers than on consumers willing to pay for the services they want. This does not mean necessarily that Americans will spend fewer hours of the day in front of the TV tubes. They may spend more. But, as suggested in a recent survey by the University of Missouri School of Journalism, they will be watching more discriminatingly because the range of available programs will have been broadened immeasurably. Not only will we have cable and videodisc but also text-delivery systems. These are variously described as Teletext and Viewdata—or whatever new buzzword may be coined at any moment by the small army of promoters now working the field.

As discrete groups within the population subscribe to particular information services for home delivery, it seems all but certain that fewer Americans will share the common pool of information. Something like this process is also under way in the newspaper business. Some large papers, even today, are designing editions to suit what they assume to be the specific interests of specific readers and advertisers. By the use of computers they are expected to zone their editions for particular geographic and demographic groups within the population. As a result, the newspaper delivered to my door may differ substantially from the edition delivered to a neighbor around the corner. Again, shrinkage in the shared information pool.

The biggest change on the horizon could be the spread of home information retrieval systems, similar to those being tested in several European countries, by Knight-Ridder in Coral Gables, Florida, and by Warner-Amex Cable Communications in Columbus, Ohio. These two systems are interactive, but they differ in several respects. Warner-Amex's Qube uses cable links to lace its network of subscribers together; Knight-Ridder uses existing telephone lines. Subscribers can order news, guides to entertainment, games, classified advertisements, even information from the pages of an encyclopedia. They can use the interactive feature to buy airline or theater tickets without leaving their homes or to order merchandise from the Sears, Roebuck catalogue or groceries to be picked up later.

To the extent that home information retrieval is taking hold by substantially answering the needs of subscribers for a wide range of information services, newspapers have reason to fear a loss of subscribers, perhaps also a decline in classified advertising linage. Once again we face erosion of the common data base. The crucial question, not yet answered, is how narrowcasting information to the home is to be financed. It is possible, however, to foresee a pattern on the horizon.

Advertising now pays all the direct costs of broadcasting and no less than 70 percent of the cost of newspapers. With the advent of cable, Home Box Office, and other supplementary services, the subscriber has been paying directly for the programs he or she chooses to receive. The richer the menu, the higher the price to the consumer. The price of a newspaper, moreover, has risen steeply with inflation and the steady increase in other production costs, including the cost of the paper on which it is printed. We pay 20 or 25 cents today for a paper that is essentially unchanged from the edition that forty years ago cost 5 cents or less. Magazines that once retailed for 25 cents cost $1.50 today. Videodiscs are not cheap, and the least expensive video recorder runs to some $600 or more.

There is the danger that sooner rather than later many Americans will be priced out of the market—debarred from the benefits promised by the new technologies because they cannot afford to pay for them. We may, in short, confront the prospect of media segregated by economic and social class: over-the-air broadcasting for the masses and the newer technologies for the classes. The affluent would be better informed than they are today; the lower orders could be even less well informed. This brings me squarely to the concern that I share with others who have qualms about the picture of the future as it is sketched by present-day technologists. I am reminded that James Madison, back in 1822, wrote to a friend: "A popular government without popular information or the means of acquiring it is but a prologue to a farce or a tragedy, or perhaps both." We can only wonder which script our age is writing.

Prophecy becomes even less rewarding when it attempts to gauge the likely impact of new communications technologies on nations poorer than our own; for example, those of the Third World. For too many years we in the West have presumed to tell these people which technologies were appropriate for them. Officials of some Third World countries were persuaded to make bad investments of this sort in the 1960s and 1970s because their eagerness to begin modernizing their own societies was in many cases not counterbalanced by a degree of technical sophistication.

I think of one official, schooled in law and philosophy, who had been persuaded to acquire for his country a satellite earth station capable of receiving satellite transmissions but incapable of transmitting. Reluctant to acknowledge that he was ill-prepared to make decisions in a matter of this sort, he naturally blamed the European supplier. No one told him, he now complains, that he had the option of acquiring a two-way earth station. Of course he attributes the bad investment to dark neocolonial designs on the part of the supplying country.

As we look ahead to the twenty-first century, incidents of this kind ought to become comparatively infrequent. The spread of technical competence is accelerating around the Third World, and we hope that the ethical standards of the sales fraternity which represents some of the multinational corporations at the supply end will also improve, whether those companies are based in Japan, the United States, France, Britain, or West Germany. The poorest of Third World countries, like the poorest of the poor here at home, will certainly not be in the market for elaborate cable systems and multimedia home consoles.

It is my guess, however, that even before the turn of the century certain of the large developing countries will have put the communications satellite to work at a job that it does supremely well—to leap over mountain barriers and penetrate trackless jungle regions at a fraction of the cost entailed in the installation of terrestrial systems. This domestic use of satellite technology, which provides low-cost links to remote areas, has been demonstrated successfully in the year-long site experiment in India. It is bound to appeal mightily to other countries that, like India, want to lace together thousands of far-flung villages for the dissemination of health and agricultural messages, news, and instruction to populations beyond the reach of the conventional media. Canada and Japan also have direct broadcast satellite projects under way which will reach out to remote areas that cannot otherwise receive reliable broadcast and telecommunications services.

I cannot vouch, of course, for the accuracy of any of these forecasts, either those offering hopeful possibilities, as just discussed, or the reverse. We are still wetting a finger to discover which way the wind blows. But I suggest that Madison had it right: farce or tragedy, the outcome will depend on the decisions we make in the next five to ten years.

I

POLITICS AND PUBLIC POLICY

In this opening section of the book national policymaking in the age of electronic communications is observed from two perspectives. It is viewed first through the eyes of U.S. Representative Timothy Wirth who heads the House Subcommittee on Telecommunications, Consumer Protection and Finance. The second viewpoint is in the form of a dissent by ecologists Amory B. and L. Hunter Lovins. The Lovinses believe that all policy decisions on important issues will be seriously flawed until information systems are taken out of the hands of technologists.

Congressman Wirth shares with Elie Abel a deep concern for the far-reaching consequences of what Abel calls "hybrid multimedia information systems." The walls of separation have broken down between discrete technologies, says Wirth, and government therefore must drastically alter the basic role that it has pursued traditionally in the field of communications. All communications policy must be reexamined in terms of the potential of the new services and new equipment that can be provided to the consumer. The goal of government, Wirth believes, should be to break down all market barriers—whether created by law, regulation, or market structure—that inhibit the introduction of new systems.

Moreover, the United States must also take a vigorous competitive position in international communications. Wirth points out that we have no coherent approach to the tough bargaining that must be undertaken if we are to confront successfully the world competition in communications during the 1980s. As a starter we must get our act together by bringing into harmony the various government agencies and departments that claim some jurisdiction over the dealings of the United States in the area of international communications.

In the opinion of Amory and Hunter Lovins the development of more sophisticated electronic systems is irrelevant. We have achieved "babble on a global scale," and the situation will get worse because the people at the top of the managerial hierarchies are not listening to what the authors call "real news." Real news is a product not of systems but of adaptive forms of social organizations called "networks." Without input from such networks or their equivalents, systems will deliver increasing amounts of information attuned to ever narrower purposes.

The Lovinses see our national effort to cope with the energy problem as the paradigm of the failed information system. The energy problem is now on its way to solution, they say—but from the bottom up, not from the top down. Because the problem is made up of billions of little pieces scattered throughout a large and diverse society, central planning tends to be more a part of the problem than of the solution. Our national experience since 1973 has "vindicated the Jeffersonians (and free-marketeers) who argued that most people are pretty smart and, given incentive and opportunity, can go far toward solving their own energy problems."

1

NEW DIRECTIONS
FOR PUBLIC POLICY

TIMOTHY E. WIRTH

In his prologue to this book Elie Abel says that the shape of communications in the twenty-first century will depend on decisions that we make in the next five to ten years. My purpose is to identify and discuss three vitally important areas in which such key decisions are going to be made in the very near future at the national policymaking level. The first group of decisions has to do with the implications of the communications revolution for education. The second concerns the way in which the federal government organizes itself to address the issues of telecommunications. The third involves the impact of communications on trade and technology. These three areas are of transcendent importance to us and to the generations that will follow.

We are in the midst of an information revolution and are fast becoming—if indeed we have not already done so—an information society. More than 50 percent of our gross national product is now attributable to activities that involve the development of data, the exchange of information, the manipulation of ideas, and the transfer of numbers. And more than 50 percent of the employment in the country is engaged, in one way or another, in these activities.

Someone once said to me that three phases in the social history of the United States can be discerned: the agricultural period of 150 years ago, in which the average person in the country was a "farmer"; the industrial society of fifty or seventy-five years ago, when the average job description in the United States was "laborer"; and the information society of 1980, in which the average job description is "clerk." That being the case, and I believe that it is, we must look at the implications for education in the 1980s and 1990s.

We are currently anticipating an expanding job market that will be ever more dependent on individuals who can manipulate sophisticated ideas and numbers. It is precisely at this point, however, when the demand is increasing for more technically skilled workers to enter the job force, that our nation's educational achievement is moving in the opposite direction. For the first time in our history the generation graduating from high school is less literate than its parents. Use whatever measure you wish: ACT scores or College Board scores, the results of New York State Regents examinations or national examinations, or measurements of the national literacy rate. The fact is that the ability of the average student graduating from high school to manipulate ideas and use numbers is decreasing. Just as the information society is calling for precisely these skills, we are losing our capacity to fill the need. This is challenge number one—a very big one.

Challenge number two is organizing our governmental bodies in a way that will enable them to examine public policy effectively at the federal level.

Historically, Congress and the Federal Communications Commission have tended to look at communications in terms of scarcity and to act solely as referees among various competing technologies; for example, a piece of the wavelength spectrum was to be maintained for radio and a piece for other uses, with government acting as referee to determine how much would be allotted for each use. We also refereed the markets allocated to television and other new technologies, including cable, which was not a broadcast medium, as they came along. The ultimate effect is that we have ended up by responding one way toward satellites, another way toward common carriers, and still another way toward computers—in the last case by doing nothing at all. We have treated each technology as a discrete area, an approach that, as of 1981, is demonstrably no longer viable. The distinctions that once existed among the various discrete technologies are rapidly vanishing; for example, entertainment and information both come into the home through coaxial cables, and telephones are used to send and receive data as well as voice.

Because of these vast changes, government must cease to consider itself as a referee and must approach the communications world in a different way. We must set a new framework for communications policy. Instead of acting as a referee, Congress and the FCC should turn the equation on its head. We should tell all interest groups that we will be looking at communications from the perspective of the consuming public—both individual consumers and business users of telecommunications systems. We should concern ourselves with the vast array of potentially available services and products. The individual consumer sitting in his or her living room does not care whether it is broadcasting, cable, satellite transmission, or telephone wire that is bringing a signal to the house. What matters is that the signal is there and that the product it is transmitting responds to the consumer's needs.

In examining the potential of new services and new equipment, we need to concentrate on three questions. First, what kind of laws inhibit the delivery of these services and equipment? Second, what kind of regulations are acting as barriers? Third, what kind of market structures may be inhibiting the delivery of new products or services?

This second challenge, the removal of barriers to the delivery of new products and services, is a challenge that is brought to us by the rapid change in technology.

The final policy area that must be examined closely in the 1980s is that of international trade as it intersects communications technology.

Whenever international negotiations are required, the United States must deal with the PTTs—the national postal, telephone, and telegraph companies of France, Germany, Britain, and other countries—as single centralized entities. There is no similar instrumentality in this country. When we go into negotiation, the FCC claims jurisdiction, as does the U.S. Department of State and the U.S. Department of Commerce. We therefore have no coherent approach to the tough international bargaining that we must undertake in the 1980s if we are to confront world competition in communications.

The Japanese have established a strong consortium between government and private business in developing the next generation of computers. The French are doing precisely the same thing in telephony by developing more sophisticated and modern national enterprises in domestic and international markets. How do we organize ourselves to take on the challenge of trade in communications or, for that matter, any other high technology? Solar energy is another good example. Under the Reagan Administration the government has decided to cut our investment in solar energy research and development by almost 75 percent, just at the time when the Japanese, French, Germans, and Italians have undertaken major national efforts to transform solar energy into electricity. Thus the issue that we face is: How do we organize ourselves within the government to help the United States continue to compete most effectively and become a greater force in the new world market?

There is nothing easy about the agenda presented.

In the area of education we are faced with a plethora of interest groups that do not wish to be concerned with the kind of change that is going to be necessary. Nor have we the kind of giant figures that we had in the 1950s and early 1960s: a James Bryant Conant who could challenge the American public education system or John W. Gardner who was able to develop his critique of higher education while at the Carnegie Corporation before going to the (then) U.S. Department of Health, Education, and Welfare. Not only do we lack that kind of figure but there is also tremendous resistance to any national examination of what is going on within our school system.

Similarly, in the area of public policy we are surrounded by interest groups that do not want the kind of change discussed in this chapter. There is resistance to examining the way in which we organize ourselves in Congress or at the FCC in relation to communications policy. These groups would have us continue to act as referee, to protect their vested interests by not adopting the user-oriented approach some of us favor.

In the area of trade and public policy there is increasing commitment to the idea that if only we can unleash the genius of the private sector all will be well. But even as our private sector is faced with the Japanese and other well-organized competitors, we lack the mechanisms that would enable us to make the long-term leap and plan six to ten years ahead as our international competitors are doing.

These are three extremely difficult challenges, but they must be met in this decade if we are going to meet the promise of the communications revolution for the twenty-first century. And because these are difficult challenges they are extremely unpopular ones for politicians to take on.

Inevitably, however, much of this will be fought out in the political arena. The best way I can summarize is by quoting Mark Twain, who told us "Always do the right thing. You will gratify some people and astonish the rest." I believe we have the possibility of doing that in the communications arena.

2

THE WRITING
ON THE WALL

AMORY B. LOVINS
L. HUNTER LOVINS

A few years ago one of the world's best equipped professional organizations for receiving and interpreting communications, the U.S. Central Intelligence Agency, failed to anticipate the overthrow of the Shah of Iran. The analysts were apparently so preoccupied with fancy electronic interceptions that they neglected to read the writing on the wall—done with spray cans. Attuned to sophisticated media and high officialdom, the agency missed the message in the streets.

This lapse is especially remarkable because the CIA reportedly allows its field analysts—not their superiors—to judge whether their findings warrant the director's attention. In contrast, the conventional hierarchies in which most managers work are simply (in Kenneth Boulding's phrase) an ordered arrangement of wastebaskets—a device for *preventing* information from

---

The authors wish to thank their colleague Patrick Heffernan for his guidance in the preparation of this chapter.

reaching the executive. Hierarchies efficiently filter out whatever the mid-level bureaucrats do not want to hear or do not consider important; their promotions, in turn, may depend on filtering the most information and making the fewest waves.

A countervailing effort to pick signals out of noise becomes ever more important in a communications revolution which vastly increases the flux of information bombarding us but makes us no more discriminating. Like Ambrose Bierce's telescope—"a device having a relation to the eye similar to that of the telephone to the ear, enabling distant objects to plague us with a multitude of unwanted details"—modern communications technology increases the noise at least as much as the signal. It demands more attention without revealing what deserves attention and it plugs us involuntarily into babble on a global scale when some of us were already adequately employed just interpreting the babble of our own familiar region.[1]

Any technology has side effects—in Garrett Hardin's definition, "consequences we didn't think of, the existence of which we will deny as long as possible." These side effects are particular to a social context. In a managerial hierarchy delivering more information attuned to narrower purposes can make the hierarchy do even better what it does worst. The technology can heighten dependence on its priests, on fallible and sometimes malicious people, and on unexpectedly vulnerable machines. It can erode privacy and further empower the State over the individual. Worse still, although communications designed by and for high technologists tend to disenfranchise and alienate the increasingly restive "techno-peasants,"[2] that is, people who feel their lives are run by remote and unaccountable technocrats, those same specialized technologies can simultaneously isolate technologists from the context of social change, making them ever more fit to manage a world that no longer exists. While building up social tensions as inexorably as misspent oil wealth in Iran, the communications adventure can anesthetize its victims by submerging wisdom and warnings beneath a surfeit of "carbonated thought."[3]

It is difficult, Niels Bohr reminds us, to make predictions, especially about the future. We hope, however, to call to your attention some disturbing bits of commonly overlooked writing on the walls that we believe will make the twenty-first-century world, in which supposedly we shall all be communicating, utterly different from a cozy technological utopia. We make no claim to a gift of original prophecy. Nothing we shall tell you is hard to find out if you know what walls to read. If you had not previously noted their significance, however, we invite you to ask yourselves why. The great Sufi, Mullah Nasruddin, once remarked that the reason the moon is more valuable than the sun is that it shines at night when we need the light more. Likewise, our own culture tells of the drunkard who looked for his lost keys under the streetlamp not because he lost them there but because that is where the light was. Just so, communications technologies that illuminate some areas can by contrast darken others perhaps more important. This shadow side of the communications revolution is at root a problem neither of hardware nor of software but rather of *purpose:* Who has an important message, whether you know the messengers, and whether you want to listen to them. A culture that *wants* to ignore those distressing spray-painted graffiti can end up with communications fatally irrelevant to its needs. This is not a side effect of new communications technologies so much as their inherent failure, when used in a particular social context, to deliver particular kinds of information; and it is on this flaw of omission that this chapter concentrates.

ENERGY POLICY
Technology Is the Answer
(But What Was the Question?)

Nowhere is this omission—the failure of a communications system to deliver relevant messages distinguished from mounds of meaningless data—more obvious than in energy policy, our main area of professional activity. It has taken ten years for the national energy debate to shift from the arena in which multibillion-dollar supply technologies battle, like giant dinosaurs, for engineering supremacy to a serious examination of the nature of the energy problem. It was discovered only a few years ago, and received into conventional wisdom only within the last year, that there is no "demand for energy" *per se;* nobody wants raw kilowatt-hours or barrels of sticky black goo. The need is rather for delivered *services* such as comfort, light, mobility, and the ability to make steel or bake bread. The energy problem, accordingly, is not simply where to get more energy, of any kind, from any source, at any price, but rather how to supply just the *amount, type and source of energy that will provide each desired service at least cost,* which you may construe either as social cost or, as in our analysis, private internal cost. This view of the problem—consistent with both common sense and market signals—has four important implications lying in ambush for any who do not heed the signals.[4]

First, most current energy supplies and virtually all marginal energy supplies are economically uncompetitive with technical measures to raise energy productivity without changing lifestyles. We are in the position of someone who, unable to fill the bathtub because the water keeps running out, has just discovered that plugs are cheaper than enlarged water heaters. "Technical fixes" are now available, for example, to double the practical energy efficiency of industrial motors or jet aircraft, treble that of lights, quadruple that of household appliances, quintuple that of cars, and increase that of buildings by ten- to a hundredfold or more, all at costs well below those of increasing the energy supply or, in most cases, of continuing to buy the energy we are buying now. So underinvested is our economy in efficient energy use that, if end-users of energy services are permitted to choose the least-cost options in the marketplace the energy needed to provide undiminished services will *decrease* severalfold over the next few decades. (That decrease has already begun.) Further, the energy sector, far from driving inflation, can then become a net exporter of capital to the rest of the economy because economically efficient energy use will *decrease* total energy costs as a fraction of GNP. Energy efficiency, then, far from being inimical to material growth, is an essential means of obtaining it—at bargain prices.

Second, although electrical supply receives the bulk of national investment in energy research, there is no market for additional electricity (at least in the industrialized countries) because electricity is an extremely expensive form of energy. Its marginal delivered price (about 8¢/kWh [1980 $$]) is equivalent to buying the heat content of oil costing $130 per barrel—four times the early 1981 OPEC oil price. The special uses that require and can economically justify such costly energy make up only 8 percent of all delivered energy needs in the United States and are already filled up twice over by power stations we have today. Debating the kind of new power station to buy is like shopping for the best buy in brandy to burn in your car or the best buy in Chippendales to burn in your stove. New power stations must compete not only with each other (around 8¢/kWh delivered) and with other generation alternatives now on the market (cogeneration, microhydro, wind, etc., at about 3–5¢/kWh) but also with efficiency improvements (about 0–1¢/kWh). Those are sufficient *by themselves* to displace probably all the thermal generating plants now operating. Because most of these options are cheaper than the *running costs alone* for, say, a new nuclear power plant, such a plant is so uneconomical that if you had just built one you would save the country money by writing it off, even at public expense, and never operating it! This may seem astounding; yet to anyone who has read the signals of electricity demand, price, and productivity it is just the unsurprising workings of the market.

Third, as the Harvard Business School energy study found,[5] the cheapest sources to meet our energy needs sustainably are (after efficiency improvements) the many appropriate renewable sources now in or entering commercial service. Currently 58 percent of our energy needs are for heat, 34 percent for vehicular liquid fuels, and 8 percent for electricity. With careful shopping and matching of each source in scale and quality to its task the "soft technologies" can meet these needs more cheaply than their nonrenewable competitors. Though not cheap, they are cheaper in capital cost, cheaper still in working capital, and several times cheaper in the delivered energy-service price than are synfuel plants or (costliest of all) the new power stations that we would have to build instead. Having never made this comparison nor allowed a competitive marketplace to make it, our nation has taken these options in reverse order, worst buys first. This is especially regrettable because the immense range of presently available renewable sources turns out, on analysis, to be sufficient to meet virtually all long-term energy needs for every industrial country studied so far—including not only the United States and Canada but such tough cases as France, Britain, West Germany, Denmark, Sweden, and Japan. Although soft technologies are neither instant nor easy, they do seem in practice to be faster and easier than their alternatives.

Fourth, the energy problem *is* being rapidly solved—from the bottom up, not from the top down; Washington will be the last to know. It has to be that way because the energy problem is made of billions of little pieces scattered throughout a large and diverse society; hence central planning tends to be more a part of the problem than part of the solution. Despite egregious market imperfections that stifle opportunity for rational choice and suppress incentives beneath energy tax and price subsidies exceeding $100 billion per year, our national experience since 1973 has soundly vindicated the Jeffersonians (and free-marketeers) who argued that most people are pretty smart and, given incentive and opportunity, can go far toward solving their own energy problems.

In 1979, for example, some 98 percent of the economic growth in the United States was fueled by energy savings, only 2 percent by all supply expansions combined: Millions of individual actions, people trying to save energy to save money, outpaced the centrally planned supply programs by more than fifty to one. Another example: Nuclear power has been under way for some thirty years with more than $40 billion in direct federal subsidies, whereas wood-burning has grown mainly in the last five years with no subsidies; yet in 1980 the United States got about twice as much delivered energy from wood as from nuclear power. With or without a truly competitive marketplace, thousands of communities and millions of individuals are identifying and grasping their local energy opportunities because they cannot afford to do anything else.

ENERGY PRAGMATISM

These conclusions are not merely theoretical. One of their practical lessons is that the United States could eliminate oil imports during the 1980s just by straightforward measures that would pay back their costs in a few years. The prescription is distressingly simple: Stop living in sieves and stop driving Petropigs. Basic weatherization of buildings could save upward of two-fifths of the entire 1980 U.S. rate of net oil imports by about 1990 at an average cost equivalent to buying oil at 15 cents a gallon. The other three-fifths of the oil imports could be saved by scrapping gas-guzzlers faster. At present their low trade-in value makes them trickle down to poor people who can least afford to run or replace them. But, compared with building synfuel plants, it would save oil much faster and be cheaper to use part of the same money to pay more than half the cost of *giving* people 40 + -miles-per-gallon cars—provided they would scrap their Brontomobiles to get them off the road. A turbocharged diesel Rabbit today gets sixty-four miles per gallon; Volkswagen has already tested a more advanced version with a composite rating of eighty to ninety miles per gallon. If Detroit retooled in one giant leapfrog to produce a fifty-mile-per-gallon fleet of cars and light trucks (twenty to forty miles per gallon worse than the state of the art) and if extra retooling costs as implausibly high as $100 billion were spread over that new fleet, then that $100 billion, at the present price of gasoline, would

pay for itself in the short span of *fourteen months*. If only we had an economically conservative Administration that read and obeyed the clear signals
of the marketplace!

A final practical lesson from that marketplace is that the electric utilities,
having for so long refused to hear the message essential to their survival,
now generally have grave financial problems. Their cash flow is inherently
unstable (and, whether they are regulated perfectly, badly, or not at all, they
will go broke if they keep on building power plants).[6] Demand may also
be so sensitive to price that higher prices actually *reduce* long-run revenues.
It is now becoming clear, however, that the problems of the utilities are not
merely fiscal but also *economic* and of a fundamental character. Roger Sant,
director of the Mellon Institute's Energy Productivity Center, showed in his
analysis of "The Least-Cost Energy Strategy"[7] that at rolled-in 1978 prices
43 percent of the electricity sold in the United States was uncompetitive
with efficiency improvements. Our analysis suggests—not to Sant's surprise—that at marginal prices *all* thermally generated electricity is uncompetitive. But those thermal power plants—more than $100 billion net worth
of them—are already built, are eating interest voraciously, and are now for
the first time being exposed to competition in an end-users' market. The
Public Utilities Regulatory Policy Act of 1978 (PURPA) goes a long way
toward creating a competitive market in the generation of electricity. Under
PURPA you can build your own generator in your factory or your backyard
or—as cheap solar cells arrive in the next few years—on your roof, and the
utilities must buy back your power and pay you the money it saves them.
You can do this *without limit*. The more power you generate and sell
back—perhaps via the entrepreneurs who are already springing up as electricity brokers—the less money the utility earns and the more overcapacity
it has. The utility must then charge higher rates. But this increases your
incentive to generate even more. Where this positive feedback loop ends,
we suspect—even if PURPA were repealed—is in the obsolescence, over
the next ten to twenty years, of more than $100 billion in thermal generating
plants. We thought, and some utilities agreed, that we had a pretty good
idea how to keep them solvent. Now we are not so sure. Utilities that didn't
read the market signals are being rapidly projected into an unregulated,
competitive environment in which they are fundamentally incapable of competing. This makes us even more worried about how pervasively utility assets
of dubious long-term worth are built into the base of a highly leveraged
financial pyramid. The desperate fiscal problems of many utilities (some big

ones may go visibly belly-up as early as this year) thus risk the whole capital structure of the United States—assuming that other vulnerabilities, such as the ability of electronic funds transfer, by mistake or fraud, to make billions of dollars disappear instantaneously, don't catch up with us first.

SURPRISES

Little of this should come as a surprise to the reader, but some of it may. Much of the writing on the wall has long been audible as a background murmur of falling utility bond ratings and common stock values, increasing federal subsidies, collapsing plant orders, and other messages of market failure. But only those who extract the message from the murmur know that solutions commonly proposed, such as higher prices or higher subsidies, are likely to make the utilities worse off. Indeed, it is common nowadays that the cause of problems is prior solutions. The energy problem can be temporarily "solved" by making it into someone else's problem—one of financial stability, or proliferation, or equity, or soil fertility, or climatic change, or water. Specialization, of precisely the kind that modern communications can bring to a pinnacle of greatest sharpness in least scope, thus defines away problems by shoving them into an adjacent pigeonhole.

Worse still, energy policy experts who spend their professional lives coping with the fallout from a singular event in 1973 often cheerfully go on, like many managers, to assume a surprise-free future. Such a future is unlikely. In 1974 one of us (ABL) made a list of the twenty most likely surprises in energy policy over the next couple of decades. Near the top of the list were "a major reactor accident" and "a revolution in Iran." (We're still looking for that list.) Number twenty, of which no examples could be given, was "surprises we haven't thought of yet." Those may be individually unlikely, but there are so many of them that they are collectively all but certain to happen.

Water policy, for example, is sending out the same ominous signals that energy did in the 1960s. This is not surprising because it embodies all the same mistakes. We are trying to supply more water before we make efficient use of the water we already have. We are treating all uses of water as alike rather than seeking the quality of water that will do each task cheapest. (The analogy to heating houses with electricity is flushing toilets with drinking-quality water.) We are seeking illusory economies of large scale in water and sewage systems, ignoring compensatory diseconomies. We are pricing water at a tenth or a hundredth of its replacement cost. We are foreclosing alternative options, having already poisoned more than a third of our dispersed waterwells. Much of our farming system is built on unsustainable mining of groundwater. The Ogallala Aquifer underlying the High Plains states is being drawn down by one to three meters per year and recharged less than one centimeter per year. It is half gone. During just the four dry months of the year more water is pumped up from this underground formation than the full *annual* flow of the Colorado River through the Grand Canyon.

Analogously, most of our farming and forestry is based on unsustainable mining of *soil*. Our farmland is losing an average of twenty to fifty tons of topsoil per hectare per year—many times the maximum rate of soil formation and more than we were losing at the height of the Dust Bowl. A dump-truck load of topsoil is washed away in the Mississippi River every *second*. In Iowa, one of the world's richest farming areas, a third of the topsoil is already gone. These losses to wind and water erosion, temporarily hidden by massive chemotherapy, do not include the soil that is burned out, compacted, chemically sterilized, or so stripped of its diverse biotic communities that it can no longer cycle nutrients effectively. These are some of the hidden costs of an overcapitalized, high-discount-rate, biologically illiterate style of agribusiness that is headed for bankruptcy as inexorably as the utilities.

How do these soil and water problems look in combination? Two-fifths of our feedlot cattle are grown on grain irrigated with Ogallala water. To grow enough corn to add enough weight on a feedlot steer to put an extra pound of meat on your table consumes 100 pounds of lost, eroded topsoil and more than 8000 pounds of mined, unrecharged groundwater.[8] Enjoy your hamburger—or your corn-based ethanol. The interactions of energy with food and water problems make the tangle worse: Strip mines destroy aquifers and farms to run water-intensive synfuel and power plants that enable farmers to pump up even deeper water—the plants themselves cooled with groundwater. No wonder Wyoming ranchers are starting to shoot the surveyors.

Climate, on which agriculture depends, has been extraordinarily favorable in recent decades but is becoming rapidly more erratic. Such well-known threats to climate stability as carbon dioxide and particulate emissions from burning fossil fuels may soon be joined by deforestation, Siberian river diversions, nitrogen fertilizer, and krypton-85 emissions from nuclear plants. Some climatologists think a single oil spill in the Beaufort Sea (where drilling has begun) could melt the Arctic sea-ice and irreversibly change the weather patterns, at least in the Northern Hemisphere.

Even subtler and more devastating surprises may be on the way in several areas of applied biology. The narrowing of the genetic base, especially for crops, is truly alarming. Some agrichemical companies are trying to accelerate and institutionalize it by buying up and acquiring legal monopolies over primitive and unhybridized germ plasm. Biocides are evolving potent new crop pests, just as antibiotics may be selecting new pathogens to prey on dense human populations. Genetic engineering is thoroughly out of control—a fit technology, as Robert Sinsheimer reminds us, for a wise, farseeing, and incorruptible people. Our society is proving to be incapable of controlling the disposition of more than a thousand new substances introduced each year, of which we have no evolutionary experience—perhaps a direct route, in David Brower's phrase, to "quicker life through chemistry."

Some years ago the World Health Organization attacked mosquitoes in Borneo with good intentions and plentiful DDT. Soon people's houses began to fall down. The DDT had killed a parasitic wasp which had controlled house-eating caterpillars. Then the cats started to die. They had accumulated lethal doses of DDT by eating geckos which had eaten poisoned caterpillars. Without the cats, the rats flourished. Faced with sylvatic plague, the WHO had to parachute cats into Borneo. Perhaps, as René Dubos suggested, the worst environmental effects are the ones we haven't yet discovered, manifestations of biological complexity that defy our simplistic mechanical intuition.

All these niceties of four billion years' biological design experience, however, may come to an abrupt end. The Hopi elders, looking toward the twenty-first century after Christ—or the sixth decade after Trinity—prophesied that a gourd of ashes will be scattered on the earth. Most students of nuclear warfare consider this quite probable. At Hiroshima this nation kindled in the sky a small sun of such intensity that bones fused to roof tiles and the granite steps of a bank were transformed into different minerals—save where the form of a human being sitting on the steps left forever a shadow in the stone. Today's nuclear arsenals are more powerful than a million Hiroshimas. They grow daily in number, deadliness, and dispersion. Any *one* of America's thirty-one essentially invulnerable Poseidon submarines has the warhead capacity to incinerate all 218 Soviet cities of more than 100,000 people. Nuclear bombs can now be delivered not only by ICBM and cruise missiles but also by fishing boat, ox cart, and United Parcel Service. A modern bomb with the explosive power of several of those used in World War II will fit neatly under your bed. The theoretical foundations of the strategic deterrence doctrine are thus crumbling fast. If New York disappeared tomorrow, we might have no idea whom to retaliate against.

Carl-Friedrich von Weizsaecker points out that just as artillery made city walls and the city-state obsolete so nuclear bombs may have made the nation-state and the institution of war obsolete. One way or another, that truth will be borne in on us by the early twenty-first century. The current Administration seems determined that it will be sooner. Decisions being made by the United States this year on grounds of pure venality will make virtually inevitable, a few decades from now, the annual flow of tens of thousands of bombs' worth of plutonium as an item of commerce within the same international community that has never been able to stop the heroin traffic.

READING THE WRITING ON THE WALL BEFORE YOUR BACK IS AGAINST IT

We have not tried to provide you with an encyclopedia of surprises but only a bit of selected advance notice. That is essential, for even though solutions *are* known or are being rapidly developed for most of the problems we have noted, implementing those solutions will take decades of hard work and a good deal of trial and error. Without foresight, we face a simultaneous onslaught by dozens of problems, any one of which would tax our scarce resources, including the attention of gifted managers.

If you have read attentively in and between the lines of *Science* and *The Wall Street Journal,* little of what we have sketched will be news to you. On the other hand, we could have told you about it several years before it appeared there. Why didn't we? Because we were talking to other people and, most likely, you were listening to other people. Techno-peasants (or, in Amory's case, apostate techno-twits) don't talk enough to chief executive officers. CEOs tend to read *Fortune,* not *Co-Evolution Quarterly, Power*

Engineering, not *Soft Energy Notes.* The real news about energy, like other resource problems, has been coming not from Exxon or the U.S. Department of Energy but from the New Alchemy Institute and the Franklin County Energy Project. Not unlike the spread of appropriate technologies in China, where news travels by word of mouth at the speed of a person walking eight hours a day, the spread of U.S. grassroots news, though speeded by telephones and photocopiers, has depended on a *social* process quite different from that emphasized in this book. The news we bring you is the product and province not of hierarchies but of a far more powerful and adaptive form of social organization—networks.

Modern communications media can bring you news and enrich a network. But, unless you pay even greater attention to who has news worth hearing than to which technology is delivering it, these same media can also deny you news and impoverish a hierarchy. Instantaneous communication does not much promote and may obscure the judgment and insight that are the hallmarks of information passed by personal contact. The getting of real news—the sort one can learn from minstrels (and perhaps from teleconferencing)—is not a mechanical act so much as a social process, and the social precondition for it is *community.*

Through greater individual responsibility and, we suspect, more sensitive antennae and hunch-detectors on the part of local managers, multinational corporations flourished in the days of the East India Company when the fastest communication was by clipper ship. A rich commercial fabric spread over three continents in the days of Egypt and Babylon when communication was slower still. If today we use our shiny new tools in ways that glorify their limitations—if we seek to get our news from machines rather than from people—then our tools may enslave and befuddle us more than they serve us. It would be all too easy to spread darkness with the speed of light.

We have not read from the wall merely as a salutary caution against the cornucopian notion that all problems are simultaneously soluble in sufficiently large amounts of money. Nor did we want to depress you with apocalyptic visions. If we had, we would have said more about global poverty, tyranny, inequity, and a half-trillion dollars in uncollectible debts. Compared to the problems we left out, those we did mention are largely soluble and, in fact, already have conceptual solutions well in hand and implementation is under way. Nor, finally, were our examples only to remind you that things are happening in the streets—of which we refrained from mentioning some of the most important—that will wholly transform before the twenty-first century who is communicating what to whom, regardless of the technical means by which it is done. Our overriding motive was rather to urge that each of us look less at how the decision-making capabilities of the manager will be enhanced by technology and more at how we can better understand the world and especially other people.

Those dazzling machines will not tell you what you need to know because the information you need isn't going into them. Most of the people who have that information aren't, and won't be, part of the organizational patterns your machines occupy. We know of people and groups, unquestionably on the cutting edge of developing practical solutions to the problems we have described, who have deliberately withheld their results from the computer data bases now being assembled for "global problem solving." Some very knowledgeable networkers won't even talk to an organization but only to individuals. And, even if your input information did not have these crucial gaps, you would probably lack the means to re-extract the signal from the overabundant noise with which it had been mixed.

By greater reliance on such double-edged tools, then, you are likely to become better informed and worse advised, knowing more—indeed, far too much—about things that matter less. If we all want to do better than merely execute disastrously misguided policies with great technical efficiency, we had better pay a lot more attention to the things that our nifty machines, and their dependents, will, if unaided, be the last to discover.

Notes

1. Paul Ylvasaker defined a region as "an area safely larger than the one whose problems we last failed to solve."

2. *The Techno/Peasant Survival Manual,* Bantam, New York, 1980 (a Print Project Book) appears to be the first popularizer of this term.

3. This phrase is due to our friend Peter Johnson.

4. Full documentation is available in our technical publications, most recently *Energy/War: Breaking the Nuclear Link,* Friends of the Earth, San Francisco, 1980, and in the periodical *Soft Energy Notes.*

5. Roger Stobaugh and Daniel Yergin, Eds., *Energy Future: The Report of the Harvard Business School Energy Project,* Random House, New York, 1979.

6. For details see Amory B. Lovins, "Electric Utility Investments: *Excelsior* or Confetti?", E.F. Hutton conference paper, March 1979, reprinted in *Journal of Busi-*

ness Administration (Vancouver) 12,1 (1981); see also his speech, *Energy Efficiency and the Utilities: New Directions,* California PUC, San Francisco, 1980, pp. 168–178.

7. Energy Productivity Center, Mellon Institute, 1980; *Harvard Business Review,* 6 (May–June 1980).

8. This estimate is due to Wes Jackson, The Land Institute: see his seminal *New Roots for Agriculture,* Friends of the Earth, San Francisco, 1980.

II

THE OUTLOOK
FOR THE MEDIA

This section takes up the problems created by the new diversity in media channels and programming brought about by the new technologies. Because new media do not replace the old but assume new functions as technology proceeds, the history of communications is essentially one of "more." This has intensified the personal-public polarization identified by Julian Huxley.

In the opening chapter J. Richard Munro states that "the basic condition of the 1980s is a new multiplicity of choices and channels on such a scale that we are confronted with a qualitative change." He speaks of "a hundred specialized lanes" now made available for children, sports fans, movie buffs, ethnic and religious groups, and others. He believes that this fracturing of markets will continue.

Two other major trends are also at work, says Munro. One refers to the kind of interchange, or interaction, that is now made possible by such systems as Viewdata and Teletext. The other trend is the development of equipment that is making this interaction far easier and more comfortable for the user. "The computer-communications marriage," says Munro, "will give us amazing power—and marvelously cheap power at that—to help guide our direction and shape our destiny."

Elihu Katz is less sanguine about the ultimate effects of this diversity. He is concerned that the apportionment of individuation and integration is becoming badly skewed because of the "barrage of individuating media." By this Katz means the customized information and entertainment available through cassette players, videodiscs, and the like. The atomization of society, he notes, has been a threat since the emergence of individualism.

The antidote suggested by Katz is "media events," which he describes as live replays of historic events-in-progress that have been prescheduled and well advertised in advance. Though often called "pseudo-events," they are defended by Katz on the basis that they give people a sense of participation and historical continuity and a feeling that individual actions can make a difference in the shaping of history.

John P. Robinson concludes from his studies of the way people use their time that we have just passed an important point in the social history of this country. People now spend more time in their daily lives communicating through mass channels rather than in direct contact with their fellow humans. The data, however, do not show that a growing diversity in programming will necessarily continue to increase the overall time people spend with electronic media. It may merely shift viewing among different choices, and the choice is more likely to be made on the basis of the quality of the image than on the content of the program.

People in general have not yet assimilated the use and the possibilities of the new electronic systems in their lives, says Robinson. It may take a generation of experience before these new possibilities "have any effect on reshaping societal consciousness of the information at our disposal."

3

UP WITH THE NEW—
AND THE OLD

J. RICHARD MUNRO

If I am tentative about looking into the future, it is because I have learned the hard way. Except for selling laundry service, magazine subscriptions, and pizza pies during college, all my working life has been with Time Inc. When I joined the company in 1957, the smart money said television was well on its way to putting magazines out of business. We published four major magazines then; we publish seven today.

The fear that television would turn magazines into buggy whips had plenty of precedents. The telegraph, many had predicted, would turn the mailman into an extinct species. The telephone would finish off the telegraph, and radio would destroy the book before it, in turn, was killed off by television. Today's folklore has it that network television is about to be destroyed by cable television.

By the time I moved within Time Inc. from what is politely called the "more traditional" media into the rambunctious world of video, cable television had already been hailed year after year as the next great medium. And year after year nothing happened. In the early 1970s, in fact, when the cost of borrowing money soared to an unbelievable 9 percent, many companies, mine included, were hastily disposing of their cable TV franchises. Today many of those same companies, mine included, are fiercely scrambling for them.

A half-dozen years ago the apostles of cable predicted that thirty million households would be wired for basic cable by 1980. At the moment there are about nineteen million. On the other hand, in 1977 a top advertising agency which follows cable television closely predicted that there would be three to four million pay cable subscribers by 1981, far short of today's nine million.

So much for old wives'—or ex-futurologists'—tales.

The history of communications is essentially the story of "more." One medium seldom replaces another. Each builds on the others. Gutenberg, Morse, Marconi, Bell, and Edison all enriched the media stream without washing up any bodies on shore. In our excitement with the new, we tend to overlook the old. Let us therefore consider print before turning to the exploding world of cable TV, satellites, videodiscs, and whatever else may have developed overnight.

With considerably less fanfare, print has actually adopted much of video's new technology. CRTs are as commonplace in newsrooms as typewriters. By using modern digital and packet-relay technology, reporters in trouble spots around the world can open a small black box to type and transmit their stories to newsrooms back home. With pictures added, stories then move electronically to printers, frequently by satellite. Transmissions from Time Inc.'s printer in Chicago, for example, move so efficiently to our Hong Kong printer by two satellite services that on Monday morning the international edition of *Time* hits many newsstands in Asia before the domestic edition appears in New York.

Coming advances in print technology will enable us to use less energy, waste less paper, and provide clearer, brighter pages. They will also make possible customized magazines and newspapers. *Sports Illustrated,* for example, will be able to add editorial pages on particular sports of special interest to any subscriber. *Fortune* or *Business Week* will be able to provide additional coverage of certain industries or financial markets. Magazines have already been customized for advertisers in geographical and demographic editions. Next comes the subscriber's turn.

Wary as I am of prophecy, I predict that in the coming years the printing industry will utilize new technologies as important to it as frequency modulation became to radio. That, of course, will change newspapers and books as much as magazines. The print media, under the spur of video competition and applying the whip of the new electronics, will enjoy its own new wave.

But the marketplace, as well as technology, determines the future. In the next ten years college-educated Americans will increase about two and one-half times faster than the overall population. Households with constant dollar incomes of $25,000 or more will grow relatively faster. Those families are and will be readers of the printed page. So will their children and grand-children.

It has become a cliché to observe that we have moved from postindustrial society into the information age, which is really nothing more than a con-venient way of pointing out that a majority of working men and women today are employed in producing, processing, storing, and transmitting in-formation. We passed the 50 percent point for information-connected work-ers only recently, yet at least one expert believes they will rise to 85 percent of the work force within ten years.

The information age results, of course, from the marriage of computers and telecommunications. Their offspring, young as they are, have already changed our lives and our media, the entertainment we choose, the knowl-edge we seek, the information available to us. It would waste the reader's time to describe those offspring in detail. Suffice it to say that the United States will have thirty or forty or fifty million wired homes in 1990 or 2010. Satellite TV, direct broadcast satellites, videodiscs, and facsimile machines will be features of the entertainment and information center that could become the centerpiece of the American home. All this has been described with great excitement in *The Wall St. Journal*, *The New York Times*, *Quest*, *Playboy*, *Time*, *Newsweek*, and undoubtedly *The National Enquirer*. What is important is to identify the major trends that underlie the vast changes that are taking place and to assess the directions in which they may take us.

MULTIPLICITY, INTERACTION, DIGITIZATION

The basic condition of the 1980s is a new multiplicity of choices and channels on such a scale that we are confronted with a qualitative change; for example, since television came of age, we have lived with three networks. A fourth network was talked about for many years. Now suddenly cable television provides not just a fourth but the likelihood of forty or even hundreds of networks. The first three resembled one another; the next ninety-seven will be, or at least have the opportunity to be, quite different. Cable, especially pay cable, has demonstrated that a great hunger exists not only for more but for difference.

The wired highway across the United States, so long heralded and now a reality, will have a hundred lanes—some fast, others slow, some a hard pull uphill for serious viewers, others a razzle-dazzle downhill slide for entertainment fanciers. We already have specialized lanes for children, sports fans, movie buffs, ethnic and religious groups, and so on. We will see those ethnic groups divided into subgroups, new culture channels divided into special cultural interests. We will witness the birth of new kinds of entertainment, new art forms, abstract entertainments, and an entertainment lane or two, I am afraid, that will be mislabeled "adult."

Some of these new lanes, no one really knows how many, will be used for information as distinguished from entertainment. Stock market and weather reports, continuous news, and Congress in session are already familiar. Almost limitless information will become available from tie-ins to immense data banks, accessible on demand. The television screen will become an information terminal able to provide endless streams of videotext, the electronic distributor of news and information.

What we don't know is how many people want it. Many companies, again including mine, are attempting to find out. Time Inc. expects to test a twenty-four-hour daily textual and graphic information service for in-home use before the end of 1981. This service will be similar to Teletext, essentially a one-way interactive retrieval system. The more advanced brother of Teletext, known as Viewdata, differs by being fully interactive in two-way communication.

Already something of a buzzword, "interaction" is the second basic trend likely to affect much of what we will see on TV screens or flat TV walls. Using Viewdata technology, the viewer can not only tap data banks or other information sources but also talk back to the computer-connected screen, instruct it, or ask it to develop earlier information in depth.

Nor is interactive video confined to information. You may one day view a movie that no one else will ever see because you yourself directed it along lines that appealed especially to you. That includes not just compressing or expanding sequences but choosing them according to your taste. You could have 100 hours of film, stills, and graphics from which to put together your own forty-minute or two-hour show, or your home computer, containing profiles of members of your family, would automatically pick certain elements and eliminate others, depending on the family member watching.

In the same way interactive capacity could make the home viewer an editor and publisher. Time Inc. is developing what we call Demand Electronic Publishing, which will enable the home viewer-reader to create his or her own magazine, to pick and choose from a sea of information photos, maps, and graphics so that some stories can be greatly expanded and others cut down or eliminated. Demand Electronic Publishing will also provide its own designs and typefaces, thus offering a range of choices to the individual viewer-editor.

A related trend of the information age is the gradual move toward a more comfortable human relationship with electronic equipment, especially the computer. One scientist working on what is often called the machine-human interface believes—these are *his* words—that "interacting with a computer can be a far richer experience than interacting with another human being." That may be a little extreme: Obviously most people are not altogether comfortable with computers, CRTs, and keyboards. How many bank customers stand endlessly in line rather than use magnetic withdrawal? How many good secretaries have thrown up their hands rather than learn to use word processors?

Time and habituation will help—witness the February 1981 Supreme Court decision to allow TV cameras into courtrooms. In part, that decision reflected our growing comfort with the TV camera. In part, it was also due to the development of new technology that makes possible smaller, less intrusive equipment.

Advances in what might be called the technology of relationships between humans and media machines are on many drawing boards or already off them. TV screens can be made touch-sensitive. Press on a particular point and that part of the screen will be enlarged. Press harder and the enlargement will be greater. Computer control by speech has also made strides. Some computers may be programmed to respond to only one particular voice. If it reveals too much tension, the user could be sent a message telling him to "relax."

No less important in the evolving machine-human video relationship will be a considerable improvement in the clarity and readability of images on TV screens and monitors. Technology will help to reduce scintillation, especially as TV switches from analog to digital transmission. That changeover will reduce the jigglies and jagglies. Special TV fonts—new kinds of letters designed by modern-day Bodonis of the video screen—will also help.

The last basic concept I want to mention is digitization, a way of encoding material for clearer and more accurate transmission. Digitization also retrieves information more efficiently. It reduces noise and snow and makes possible a wide range of special effects. The TV games that you perhaps play, and that your children certainly do, depend on digitization. So do the new videodiscs, which bear no more resemblance to phonograph recordings than they do to the wheel.

RIDING THE WAVE
Is It Enough?

Digitization, comfortable man-machine relationships, interactive video, and multiplicity all will be fundamental to the video media we will be using in the year 2001 and beyond. New media do not replace old, nor do they work as first intended. Gutenberg considered his movable type primarily a way of making the Holy Bible more broadly available. It did—and paved the way for the Reformation. Marconi thought his wireless was a better means of point-to-point communication. What he gave us was the radio. As for television, it was generally conceived as a medium for visual instruction. In each case the socioeconomic system, not the technology, determined the use. Progeny of the computer-telecommunications marriage will certainly make possible many shifts in our social behavior. But they cannot guarantee which shifts will occur.

However one feels about commercial television, its tremendous effect on our lives cannot be denied. Even those critics who accuse network television of having invented cloning well before the geneticists will admit that TV played a powerful role in ending a war in Vietnam, unseating a President who used the medium badly, and electing others who used it well.

Before television, radio produced social changes no futurist could have foreseen. Coming afterward, the computer was first thought of as a kind of mechanical brain. Then vacuum tubes replaced the mechanical relays, transistors eliminated the vacuum tubes, integrated circuits superseded transistors, and the computer-on-a-chip is replacing the integrated circuit. The mechanical brain became something far different, with certain social and economic consequences.

The telephone and television, even the computer, were basically single inventions. The age of information, however, is a constellation of inventions that converges in the marriage of computers and telecommunications. That convergence could change our lives as spectacularly as the industrial age changed the age of agriculture before it—for better or for worse. I believe the change will be for the better, for two reasons.

First, mankind slipped into the industrial age with little realization of what was happening. In fact Arnold Toynbee dubbed it "the industrial age" perhaps a hundred years after it had begun. Our early warning system has alerted us to the new change in its infancy. Aware of its potential and its dangers, we may manage it without the social equivalents of child labor, dark satanic mills, and deep economic depressions.

What is more important, the direct consequences of the new media occur in the realm of the mind. The steam engine, like the printing press, replaced physical effort; the industrial age was built largely on substitutes for muscle. Today's computer-telecommunications media are, or can be, the servants of analysis and thought—highly trained professional servants, if you will, rather than the domestic ones of the industrial age.

It is rather surprising that our new information age has not produced its own philosopher. Some years ago we thought we had one in Marshall McLuhan, but the one-person movie or the one-time magazine certainly do not move us toward McLuhan's "global village"; instead of moving toward tribalization we seem to be achieving greater fragmentation.

My own favorite philosopher-futurologist is Arthur Clarke, in part because in 1945 he first described today's communications satellite. By general agreement Time Inc.'s use of the communications satellite in 1975 gave cable TV the wings that turned a struggling business into a strong industry. Arthur Clarke believes that the wave carrying us into the future has scarcely started its run. Behind us lie reefs, beneath us a great wave arching ever higher. Where, he asks, is it taking us? Clarke answers his own question: "We cannot tell . . . it is enough to ride the wave."

Elemental, evolutionary forces before which we are sometimes powerless do exist. But it is *not* enough just to ride the wave. We are not surfers on a joyride but navigators obligated to chart a course. The millennial year 2001 seems unlikely to be the millennium in the sense of happiness supreme and human perfection. If information is power, as I believe it is, the media children of the computer-telecommunications marriage will give us amazing power (and marvelously cheap power at that) to help guide our direction and shape our destiny—to dominate the wave we are riding. Together with the older media, which the new technology replenishes rather than replaces, they will entertain as well as instruct us, and make us smarter and perhaps even wiser.

4

IN DEFENSE OF
MEDIA EVENTS

ELIHU KATZ

with DANIEL DAYAN *and* PIERRE MOTYL

This chapter is concerned with media events of the kind that may be described as "the live broadcasting of history" or, alternatively, "the high holidays of mass communication." Our interest in this genre was stimulated by Sadat's dramatic visit to Jerusalem. We were soon comparing it with other "great steps for mankind," such as the moonlandings and the Pope's visit to Poland. The corpus of material we have collected—videotapes, films, research reports—relates to events that electrified a nation or the world. The reader will soon gather that we have more than the usual sympathy for this form of symbolic representation, despite the grave doubts about its integrity expressed by many scholars and critics and even despite certain reservations of our own on this score. Viewed

The authors have drawn on research and ideas from their forthcoming book on media events. The project benefits from a grant from the John and Mary R. Markle Foundation and from the Joint Committee of Trustees of The Annenberg School of Communications at the University of Southern California. The authors thank Carolyn M. Spicer for editorial assistance.

analytically, these "events" have revealing things to disclose about television and about our present society; they have a bearing on the twenty-first century as well.

We begin with an attempt to place media events in the larger context of mass communication with particular reference to the division of labor among the media—past, present, and future. We look next at the characteristics of media events, then at the alleged dysfunctions and dangers inherent in these political ceremonials, and finally at the positive functions that seem deserving to us.

INTERRELATIONS AMONG THE MEDIA

The social history of the media appears to tell a story not of displacement of one medium by another but rather of differentiation by which the media divide themselves, at any given time, among the social functions that have to be performed. When a new medium emerges and addresses itself to a particular need, the tasks of the remaining media are reshuffled.[1] Thus certain media liberate their predecessors in the sense that photography liberated painting from its commitment to realism or—to stretch the point a little—in the sense that television liberated radio for intimacy and companionship. Some media provide content for one another as McLuhan suggests. The novel became content for film, the phonograph record for radio, and the film for television. Media may also exchange functions. Thus when the telephone was first introduced, it was apparently conceived as a broadcast medium, whereas radio was thought of in terms of point-to-point communication. Obviously the media may recombine in new ways to bring about radiophoto or Teletext or any of the myriad new combinations that we may contemplate.

In the midst of these musical chairs we should not lose sight of the communications needs of individuals and societies that the media are designed to satisfy. These needs have been variously classified. Harold Lasswell, amended by Charles Wright, identifies four basic needs[2]: *surveillance*—a scanning of the environment, as if by social radar; *correlation*—the editorial effort to make sense of the information obtained; *socialization*—the transmission of values across generations; and *entertainment*. Another company of scholars has distinguished between the needs of *fantasy* and *reality*.[3]

Still another set of distinctions, which has evolved from a research tradition in which we have worked, points to the need for *information* or knowledge for *pleasure*, for a sense of *identity* and status, for *connection* to the larger society, and, alternatively, for disconnection or *escape*.[4]

Different media specialize in serving different needs. In our Israeli studies, for example, we found that the newspaper is still the medium of choice when people seek connections with community, nation, and the world, whereas film rates highest as the medium for enjoyment, followed by television and books, in that order.[5]

The different schemes for the classification of communications needs can be made to converge in several ways. One axis appears to be reality and fantasy or information and entertainment.

Another axis—of particular interest to us in this study—distinguishes between communications that serve intimacy and individuality and those that reach out to integrate large groups of people. The distinction here, if you wish, is the one already mentioned that contrasts point-to-point communication with one-to-many (broadcast) communication.

The basic needs for individuation and integration exist in all societies, albeit in differing proportions. In modern society it is obvious that certain messages and media are designed to enhance the highly specialized interests of individuals or identifiable subgroups, whereas others cut across interest groups to unite entire societies. Thus *Life* and *Look* made way for the special-interest magazine on the one hand and for the leveling medium of television on the other. These dual tendencies may be seen even in ancient societies. Harold Innis, for example, writes about media that specialize in overcoming space and media whose messages traverse time. Innis considers Roman highways, the Nile River, and writing on papyrus as the media of empire that enabled centralized regimes to keep their far-flung bureaucracies updated and under control. On the other hand, the less portable pyramids and temples, as well as chiseling on stone, were used to speak across generations and between the living and the dead.[6]

A DIGRESSION ON THE BOOK

Media technologies, like other technologies, become pliable in the hands of humans and may be pressed into the service of needs for which they are ostensibly unsuited. The flexibility of the telephone and radio have already been noted, but the book is surely our best example.[7] Ostensibly, the fragility of the book would seem to relegate it to the medium of space rather than time, to the function of intimacy and identity rather than large-scale integration and community. This is certainly the case today. People sit in their own corners reading their own books; only occasionally does a bestseller come along to offer a shared experience to a large number of people. That was not always the case.

There was a time in both Christian and Jewish tradition when the book—The Book—served as a medium of wide-scale integration. For the Jews the Bible was a portable homeland, a medium of time and space. On the Sabbath—indeed, four times each week at an appointed hour—Jewish congregations throughout the world gathered to read the same text, aloud and in public. The traditional experience of reading was collective rather than individual, and even when the people were literate the text continued to be recited and discussed. As a direct result, the Hebrew language remained alive, and metaphors and heroes were shared throughout the diaspora. Cultural creativity consisted of contributing to the interpretation of The Book, communities exchanged scholarly messages, and a vast literature was accumulated. Knowledge of The Book was an important route to status in the community, and the renown of scholars was widespread.

It is no exaggeration to say that "the portion of the week," as the lesson of Bible readings is called, played a major role in the integration of the Jewish people. It is a great irony in Israel today, where book reading is still remarkably popular, that books are the medium of personal experience, both aesthetic and cognitive, whereas the portion of the week—on Israel's one-channel television system—tells of Kojak, Quincy, and Henry VIII.

THE NEW MEDIA AND SOCIAL INTEGRATION

Although media technology is pliable, as we have seen, it certainly looks as if the future has a barrage of individuating media in store. The multiplication of television channels, the direct access to customized information and entertainment, the home cassette player, and the videodisc all are aimed at perfecting individual autonomy. People will be able to dial their own identities with ease. But what will unite them? What will offer a shared experience to a nation or to whatever political-social-cultural unit that will supersede the nation-state? The media of integration seem to be on the wane.

The atomization of society has loomed as a threat since the emergence of individualism. Modernity has cut people off from the assurance of membership in a craft, in a church, and even in the family. Modernity has created institutions in which people relate to one another only in highly segmented roles rather than as whole persons. Theorists of the mass society have remarked on the vulnerability of people to external manipulation under circumstances of loneliness, isolation, and alienation. Erich Fromm's warning of the temptation to "escape from freedom" still stands.[8] The weakening of ties to a center vested with legitimate authority is an invitation to a demagogue to use the media to reunite people. Khomeini used audio cassettes to rally his people in revolt against an authority that had lost its legitimacy in spite of its control of the most powerful and sophisticated broadcasting organization in the Third World.[9] The need for a sense of community, even in modern society, remains strong. If the political and cultural centers of the society cannot provide it, others will try.

At the moment television is our focus on the center of society. It is the medium that we all share, like the weather. We know now that the ostensible diversity of channels and programs is really quite narrow compared with the possibilities ahead. It is also clear, in spite of the criticism, that the relative independence of the communicators serves, on the whole, to safeguard freedom and to maintain high professional standards, at least in the realm of public affairs.

Yet it may be that only the live broadcast will remain on television as we know it. The real-time transmission of events as they occur and the swift correlation of news and current affairs may possibly stand a chance of competing with the new technologies. At least one country, Mexico, has been talking of experimenting with an all-live television channel to bring news and entertainment into the home in real time. It is a nice paradox that television broadcasting of the future may return to the days before videotape.

MAJOR MEDIA EVENTS

It is in this context that we wish to discuss media events. Over the past twenty years we have all joined in live broadcasts of the moonlanding, John F. Kennedy's funeral, the Apollo flights, Sadat's visit to Jerusalem, the pilgrimages of the Pope, the World Cup, the Super Bowl, the Olympics, national political conventions, presidential debates, the Watergate hearings, and the return of the hostages. We will not add "and many others"; these are very special events, and there are not—indeed, cannot be—many of them.

In our research[10] we find that these events fall into three major categories. We call them, for fun, contests, conquests, and coronations. By contests we mean presidential debates or sports events in which superplayers, evenly matched, compete by the rules. Conquests tell the story of a hero—like Sadat, the Pope, or the astronauts—doing the impossible, pushing back some frontier. Coronations are funerals, parades, academy awards, or inaugurations to celebrate heroic accomplishments or august offices.

These events call a nation together to sit down and be counted. They have caused people to dress up, rather than undress, to view television. They have brought friends and neighbors into the living room to watch together because one wants real people, not just the furniture, to share the experience. These events have animated discussion and sometimes— perhaps as in the wake of the Pope in Poland—have served to stimulate action. And the public reaction feeds back on the participants, making them feel stronger and freer.

To generalize, these broadcasts have in common that they are live relays of historic events-in-progress. They are interruptions of routine for the participants, for the public, and for the broadcasting organizations themselves. They offer the drama of simultaneity, in which a story is told whose end is still uncertain. They speak of heroic deeds. They are laden with symbols.

It should be noted that these events are by no means spontaneous. They are prescheduled and well advertised in advance. When successful, they enlist the support of social norms that "require" people to listen or view. Typically, they are presented without criticism and often with some degree of reverence by journalists who, in other circumstances, are critical and even cynical. Commercials are often suspended to honor the sanctity of the occasion or especially tailored to the event. It is important to add that most of these events would have taken place in some form even without benefit of the media: Kennedy and Mountbatten would have been buried; Sadat might have gone to Jerusalem; the astronauts would have flown to the moon. But it is also true that the media, especially television, have a major influence on the form these events actually take.

One might say that all this sounds like a good thing—a possible antidote to the atomizing tendencies of modernity, present and perhaps future. But we have been warned against such events, and before going deeper into their functions we are obliged to consider their alleged dysfunctions.

THE CRITICS OF MEDIA EVENTS

The most famous critique of media events is that of Daniel Boorstin.[11] In his book *The Image,* published in 1962, Boorstin grouped our kind of event with the interview, the panel discussion, the handout or press release, the press conference, the leak, and the glamorizing of political or diplomatic news with details of the personal thoughts and actions of the officials and politicians concerned, all of which he labeled "pseudo-events." In struggling to define these pseudo-events, he contrasts them with spontaneous events, whose reality is only minimally tampered with in reports by participants and journalists. Boorstin is ready to grant that he cannot cope with a formal definition of the real; he would probably admit that all perception is structured and selective. Yet he longs for the yesteryear of the spontaneous event that was reported immediately as it happened, unstaged, unadvertised, uncontaminated by professional producers, speechwriters, image makers, or newspapermen who think that good stories and pictures are "better" than reality.

Almost twenty years have passed since the publication of Boorstin's book. Looking back, one finds it difficult to reconstruct the kind of journalism he implies we were losing. The reporting of a train wreck is his own best example. Actually it is modern journalism itself that worries Boorstin—the frenzied editors who cabled their man in the Caribbean "you supply the pictures, we'll supply the war," or words to that effect. Since then things have become worse, Boorstin believes. Speechwriters and media advisors accompany the President everywhere. The press conference promises that news will break at the next press conference, a favorite tactic of Senator Joseph R. McCarthy. The handout describes a future event in the past tense and in language acceptable to the establishment; any departure from the prepared text then becomes new news. Interviews and off-the-record briefings give control to the newsmaker, with the journalist as his active collaborator.

Events are especially troublesome, although there is some ambivalence on this point in Boorstin. Some events—not only fires—sometimes really happen, and when the reporting is swift, verbal (broadcast), or minimally manicured by press agents or newsmen, Boorstin is satisfied. Most news events, thinks Boorstin, are staged for the press. They would not happen at all if it weren't for the presence of the media; they are hyped up dramatically to serve the interests of the newsmaker, the advertiser, or the publisher; they are carefully timed in the doing, or in the presentation, to the leisure hours of the public. The genre thrives because modern persons prefer fantasy to reality and readily consent to be fooled all of the time. Worst of all, says Boorstin, the effort to demystify the process by exposing what goes on behind the scenes only makes pseudo-events *more* attractive.

Boorstin reminds us of the Langs' classic comparison of the experience of a television audience watching General MacArthur's return to the United States in 1951 with reactions of the people who lined the streets on the route of march.[12] The TV viewers were much more emotionally affected than the people in the streets. The latter only saw the motorcade pass by, whereas the TV viewers followed the parade as it moved from the airport to the city hall, listening all the while to the narrator tell the hero's tale.

This is unreal, says Boorstin. "In the age of the pseudo-event," he writes, "it is less the simplification than the artificial complication of experience that confuses us. Whenever in the public mind a pseudo-event competes for attention with a spontaneous event in the same field, the pseudo-event will tend to dominate." The Kennedy-Nixon debates taught us nothing; but their staging—like quiz shows, with journalists as quizmasters—was enough to induce a dislike for normal political speechmaking in the largest audience ever assembled for a political broadcast. These events, adds Boorstin, are not hoaxes; they are not created by crooks but by honest people serving principles such as "the media must be fed" and "the people must be informed."

Almost thirty years before Boorstin's concept of the pseudo-event the brilliant critic Walter Benjamin wrote that the strategy of fascism is to make politics into aesthetics.[13] Four years before his desperate attempt to escape the Nazis, Benjamin was trying to warn us not only against those who dress up in uniforms to stage parades and mass rallies but against the whole enterprise of making politics dramatic. Better to realize, says Benjamin, that drama is also political. Thus, he says, the Marxist reply to fascism is to politicize art.

If Boorstin's reality is to get as close as possible to the naked thing itself—the fire, the train wreck, the parade, the presidential candidate—Benjamin's way to reality, so say his biographers, is to dig deep within the mysteries of language and sacred texts. The climax of his warning is this: "All efforts to render politics aesthetic culminate in one thing, war," and he cites the fascist aesthetic that turns men into machines, technology into art. In a final footnote Benjamin returns to the film, the subject of his essay. "In parades and monster rallies," he says, "in sports events and in war, all of which nowadays are captured by camera and sound recording, the masses are brought face to face with themselves. . . . Mass movements are usually discerned more clearly by a camera than by the naked eye." The propagandistic function of the newsreel cannot therefore be overestimated, Benjamin concludes. For all their ideological differences, Boorstin and Benjamin may not be so far apart.

But one can object to media events for quite a different reason. Instead of objecting to their unreality it is possible to argue that they are too real. This is the point made by historian Michael Confino in his essay *Historical Consciousness in Contemporary Culture*, written almost twenty years after Boorstin's *Guide to Pseudo-Events*. The problem with media events, Confino says, is not only that they improve on reality but they are embedded in the syndrome of "being there."[14] The closer you get to an event, implies Confino, the less you see of it and the less you understand it. The stuff of history is not event but process, he argues. We escape from history by trivializing it. The slightest surprise is labeled "historic" by participants and journalists trying to inflate their stories. Textbooks and journalists give events a prominence they do not deserve, and events oversimplified in the past are then misappropriated to explain events in the present.

This part of Confino's argument alludes—or so it seems—to the virtual banishment of events from the realm of social science, including academic history. Explanations of historical process are offered in terms of economic and social dynamics, not in terms of heroic deeds. Boorstin regrets this, Confino is pleased, but both agree that heroes are passé. They have been psychoanalyzed out of existence or shown to be the creatures of folk psychology. Accomplishment is so complex nowadays that we cannot revere the true heroes of modernity because the true heroes are likely to be presiding quietly over a team of scientists or explorers or statesmen rather than performing publicly as they once did. Celebrities—creatures of publicity and public relations—have taken their place, says Boorstin.

Confino would say get rid of events and heroes. Boorstin would be pleased if journalism could spare us some real events and real heroes, if only we could get rid of media events.

To reiterate, three arguments are made against media events. One, Boorstin's, holds that these events would not have happened if it were not for the active collusion, or at least the proximity, of the camera and the microphone. A second, Benjamin's, is that media events can be manipulated to whip masses into the frenzy of chauvinism. Finally, Confino's argument is that "historic" events, especially media events, divert us from a true consciousness of history. "If a tree fell in the forest and the media were not there, did the tree really fall?" Boorstin would say yes. Confino would say that single trees are irrelevant; it's what's happening to the forest that matters. If the media *were* there, Boorstin might say they would first advertise that the tree was going to fall, they would then show it being felled by a woodsman wearing an Adidas shirt and quoting the Bible, and finally they would commit an instant analysis of why it fell at an angle different from the one predicted. Confino would say that you can't see the forest for the tree, and Benjamin would say that the event of the felling of the tree is a message about the unrelenting power of the authorities and their use of fearsome technologies.

IN DEFENSE OF MEDIA EVENTS

Defenders of media events have to explain themselves: It is not enough to answer that these occasions make good stories or that they reawaken a sense of collectivity. Before reiterating the positive functions of live broadcasting of heroic deeds one must confront the critics.

The consensus of the critics—these three and many others—seems to be that mass media distance and distort reality. There is not much to disagree with here. Of course the media distance and distort reality! Lacking direct contact with much of our environment, citizens of modern industrial societies depend on the media to act as their intermediaries. There is distortion, too, influenced by patterns of ownership, media technology, professional practice, and external pressure. Much research has gone into an understanding of the rhetoric and images of the mass media and into the search for an alternative doctrine to succeed the professional credo of objectivity. By definition media "come between." This is as true for historians as for journalists.

One can agree that the real and the mediated have become hopelessly intertwined. The report of a strike—indeed the very rumor of a strike—

becomes part of the reality that is then acted on. This, in turn, becomes a part of the subsequent reportage. For this reason it is difficult even to talk about the effect of the media.

There is no point in saying that Sadat would have come anyway or that the astronauts would have gone to the moon even if they had not had live television coverage. One can imagine how it might have been without television, but the very fact that Sadat's arrival in Jerusalem was seen on television by most Israelis and Egyptians and by many millions elsewhere became, in itself, part and parcel of the reality of his arrival. One may have other sources of knowing, but most of them are also media-fed. The more complex the society, the less one learns from direct personal experience and the more one learns from intermediaries.

Having granted the disparity between reality and constructed reality, one can study the dynamics of this transformation and heed the warnings of its possibly negative effects. Media events may keep us on an exaggerated high; they may wipe out genuine heroes; they may be manipulated by those who wish to mobilize emotions in support of a regime; they may distract us from a more sophisticated view of history. All this, under certain conditions, is possible.

We are very well aware that the event itself is an editing. Its presentation is based on a choice among competing scripts even before the event has transpired. Moreover, this editing is only the first of a long series. During the event there are daily wrap-ups, then a general wrap-up; there is a selection of highlights from the news. Subsequently, there are special fea tures and in-depth analyses, then commemorative broadcasts. These successive editings displace our own decisions concerning what to remember and what to forget. The media function as our memory, a simulation of it; they transform our experience into a visual cliché.

We do not underestimate the danger or the possible misuse that can be made of this genre, but we are equally aware of its potential for good. It is to this rosier side of the picture that we now turn. In the following discussion of the more positive aspects of media events we have drawn particularly on our studies of Sadat in Jerusalem, the Pope in Poland, the moonlandings, the Kennedy funeral, and presidential debates.

Media events testify that *voluntarism* is still alive, that the deeds of human beings—especially great ones—still make a difference and are worth recording. People want to hear this and storytellers want to tell it, even if our critics are not all equally pleased. Extending this argument, one may assert that journalism is the last great refuge of events. History and the social sciences have banished them in favor of more abstract analyses of process. But the will and the deeds of individuals still count for a lot in the daily news. Boorstin might be satisfied with this state of affairs if more serious efforts were being made to separate the real from the fabricated. Confino would still find the reporting of events irrelevant and distracting. Benjamin would find them dangerous.

Media events are *integrative* because people can identify with them. Unlike the nightly news which is typically about conflict—man against man or man against nature—media events more often celebrate the overcoming of differences and the reconciliation of conflict. The circling of the moon on that Christmas Eve by Apollo 8 was a transparent effort to marry two sets of symbols. And people felt that the marriage was right. When the astronauts scrambled out onto the moon, the world was reunited, just as two nations united when the El Al staircase was locked to the door of the Egyptair jet that brought President Sadat to Israel.[15] The pageant of the Kennedy cortege caused a nation to embrace itself, just as it had during Lincoln's long, last journey.[16]

Just like public ceremonies in traditional societies, media events are *reflexive* in that they connect people not only to one another but to values and beliefs that are central to the society. Some critics will see this as manipulative and as reinforcing the status quo. The moonlandings were a celebration of American technological prowess, a dramatization of man's ability to control nature. If this is the aesthetizing of politics, it may quickly be added that Three Mile Island echoed man's deepest anxiety over losing control of nature and technology. John Kennedy's funeral or Elizabeth's coronation gave Americans and Britons an insight into the measure of their personal attachment to basic institutions in real or imagined jeopardy.[17]

Again, some critics will see this as transparent maneuvering to preserve the status quo, in which case one must look to the Watergate hearings or to the days after the Pope's visit to Poland to see how risky a game this can be. The Pope in Poland succeeded in reuniting a people with their history and with the myths and symbols from which they had been separated for thirty years. But they had to carry on by themselves after he left.

Indeed, media events have the quality of *holidays*. Like holidays, they are interruptions of daily routine that call attention to something special. They are governed by different rules of dress and food and speech; they draw on the shared sense of community associated with the occasion. For a society into which secularism has made strong inroads the sense of occasion engendered by very special media events—even the Super Bowl, but certainly the moonlandings or the welcoming of the hostages back from Iran—have a quasi-religious atmosphere. The media—participants and presenters alike—reflect this spirit. Remarking on the reverence of journalists covering the Pope's visit in the United States, Garry Wills calls it, irreverently, "falling in love with love" and "the greatest story ever told."[18]

Not all events succeed. The promoters of an event can guess wrong; the media—to the extent that they are independent of government and the establishment—may agree to treat the wrong affair as an "event." Ultimately the people will decide—in their viewing behavior, in their conversations—which events will succeed and which will fail. Many people would agree that the 1980 party conventions were events "manqué"; I'm not so sure. Cautiously, one might suggest that the people's response to events so presented may be tests of their *authenticity*. When Sadat arrived in Jerusalem, there were very few cynics left who doubted the genuineness of his peace initiative, even fewer who doubted the genuineness of the Israelis' desire for peace. Sadat and Israel confronted each other as much on television as on the streets.

Events induce *participation*. Critics like Confino believe that media events heighten the feeling of alienation from the center, making the contrast between the actions on the screen and the inaction in the living room even more vivid. But this does not seem to be the case. First of all, live broadcasting of events constitutes a focus for the expression of emotion. The risk of a misstep (diplomatic or astronautic) and the unknown outcome induce excitement. Televising the Kennedy funeral became a focus for the expression of grief; Schramm's review of the research documents this very movingly.[19] The ability to stir emotion and focus it is no small feat.

There are cognitive effects as well. Although the mass media do not so easily effect change in opinions and attitudes, media events may have the rare potential for doing so. Sadat did change the attitudes of Israelis and Americans toward Egypt, just as the Apollo mission changed European evaluations of American ability.

But beyond emotional expression and attitude change certain media events make possible genuine political participation. The spontaneous mass support for the dialogue between Begin and Sadat liberated these two men to do far more than their associates or their reference groups would otherwise have allowed them to. The Poles saw themselves politically reanimated by the Pope, and they invested their new potency in their church and their emergent labor unions, bringing it to bear on their government. The Reagan-Carter debates may have made voting for Reagan possible for many people who might otherwise not have allowed themselves to consider the possibility.

Participation is made easier because the *issues are crystallized*. Sadat confronted the Israelis with the clear-cut question, which they had been avoiding, whether they would trade territory for a gesture of peace. The Three Mile Island affair took questions of energy out of the realm of technology and bureaucracy and placed them right back on the political agenda of public opinion. Boorstin dismisses the first Kennedy-Nixon debates because they dealt with irrelevant issues, but the debates—thanks to the journalist questioners—forced the candidates to confront a number of issues that they would gladly have avoided.[20] The debates also made manifest that personality *is* a political issue.

Finally, media events are concerned with *process*. They give some insight into how things work. Thus we had a glimpse of Wilson's old dream of "open diplomacy" in the Sadat-Begin encounters. With the Pope we had some insight into the dissonance between his open gestures and his conservative doctrine. Thanks to the media and their production routines, we get some insight into how an event is put together. We see it not only from the sidelines but from front and back and even above—as in the beautiful shot of Mountbatten's bier taken from high in Westminster Abbey. We are given explanations of symbols that would otherwise pass us by. We learn about what is going on backstage while commentary and critique offer us larger perspectives into which to place the event. Confino thinks that media events of this sort distract and mislead from a true understanding of historical process. Indeed, they obviously are not the best didactic form for serious students of history, but for most of the rest of us they tangibly connect the present with the past and future, superficially perhaps, but in a way that is accessible to a very large audience.

CONCLUSION

Live media events appear to have a future, not just a past. The new media will defer to television's unique ability to bring us historic and heroic events as they are happening.

There are dangers in this, some of which we have enumerated, but they are tempered, in democratic societies at least, by institutions. The media are relatively independent of government, and not every parade or anniversary will become an accredited media event. The people are still free to view or not, and not every media event will succeed. Not all events support authority: witness Watergate and the Pope in Poland.

Elsewhere, we analyze the genre of media events more carefully and more critically. The object here was to present some of the pros and cons of showmanship of this kind within the historical and societal contexts of the late twentieth century in order to point up some of the possible implications for the future.

We have made no effort to conceal that we are sometimes carried away by the thrilling nature of media events. Indeed, it is hard to resist the memory of some of them. They evoke a sense of community and occasion, of connecting with authentic values, of participation—a feeling that perhaps one makes a difference after all.

We are not shying away from the hard questions. We have asked ourselves, "If the media are now custodians of our memory, and successive editings narrow our image of history, how shall we be able to critique their theory of the past when it is all we can remember? How shall we distinguish media historiography from other evidence of reality including our own experience?"

Still, it is a fact that when a media event truly fires, when it rings the bells of authenticity, it has a validity that cannot be exaggerated. Consider the Pope's visit to Poland. A group of dissident Polish sociologists conducted interviews throughout the country in the wake of the visit, and we have a copy of their still-unpublished and untranslated manuscript. One of the interviewers offered this assessment of the Pope's homilies as seen and heard by millions in person and over the air:

> For the first time, words began to fit the reality, the consciousness and the experience of the people who heard them. These words were being reloaded with meaning. It was as if their real semantic value was given back to them. It was like the instant of emerging from schizophrenia, when what one says publicly is what one really feels and wants. It was like a flash, as the possibility of experiencing individual and national autonomy opened up. It was the feeling of getting in charge of your own fate again. People were realizing that after all they are not powerless, that what will happen when this visit ends depends somewhat on them, that something of the future is in their hands.

Not every media event is like the Pope's visit to Poland or the first celebration of the mass over television in a Catholic country of Eastern Europe. But there may be some connection between media events and historical consciousness after all.

Notes

1. This paragraph draws inspiration from Marshall McLuhan, *Understanding Media*, McGraw-Hill, New York, 1964.

2. Harold Lasswell, "The Structure and Function of Communication in Society," 1948; Charles Wright, "Functional Analysis and Mass Communication," *Public Opinion Quarterly,* **24,** 605–620 (1960).

3. Wilbur Schramm, Jack Lyle, and Edwin Parker, *Television in the Lives of Our Children,* Stanford University Press, Stanford, 1961.

4. Jay G. Blumler and Elihu Katz, Eds., *The Uses of Mass Communication,* Sage, Beverly Hills, CA, 1974.

5. Elihu Katz and Michael Gurevitch, *The Secularization of Leisure,* Faber, London, 1973.

6. Harold Innis, *The Bias of Communication,* University of Toronto Press, Toronto, 1964.

7. Katz and Gurevitch, *Secularization of Leisure, op. cit.,* Chapter 12.

8. Erich Fromm, *Escape From Freedom,* Farrar and Rinehart, New York, 1941.

9. Majid Tehranian, "Iran: Communication, Alienation, Revolution," *Intermedia,* **7,** 6–12 (1979).

10. For an introductory statement see Elihu Katz, "Media Events: The Sense of Occasion," *Studies in Visual Communication,* **6,** 84–89 (1980).

11. Daniel Boorstin, *The Image: A Guide to Pseudo-Events in America,* Atheneum, New York, 1964.

12. Kurt and Gladys E. Lang, "The Unique Perspective of Television," in *Politics and Television,* Quadrangle, Chicago, 1968.

13. Walter Benjamin, "The Work of Art in the Age of Mechanical Reproduction," in *Illuminations,* Jonathan Cape, London, 1970 (translated from the German).

14. Michael Confino, *Historical Consciousness in Contemporary Society,* The Aranne Foundation at the University of Tel Aviv, Tel Aviv, 1980.

15. Elihu Katz, with Daniel Dayan and Pierre Motyl, "Television Diplomacy: Sadat in Jerusalem," lecture given at the Conference on World Communications: Decisions for the Eighties, The Annenberg School of Communications, University of Pennsylvania, May 1980; Tom Wolfe, *The Right Stuff,* Bantam, New York, 1980.

16. Lloyd Lewis, *Myths After Lincoln,* Harcourt Brace, New York, 1929; Wilbur Schramm, "Communication in Crisis," in Bradley Greenberg and Edwin B.

Parker, Eds., *The Kennedy Assassination and the American Public*, Stanford University Press, Stanford, 1961.

17. Edward Shils and M. Young, "The Meaning of the Coronation," in Edward Shils, Ed., *Center and Periphery*, University of Chicago Press, Chicago, 1975; Asa Briggs, "Television's Coronation," in *History of Broadcasting in the United Kingdom*, Vol. 4, Oxford University Press, London, 1979.

18. Gary Wills, "The Greatest Story Ever Told," *Columbia Journalism Review* (January–February 1980).

19. Wilbur Schramm, "Communication in Crisis," *op. cit.*

20. Sidney Kraus, *The Great Debates*, Indiana University Press, Bloomington, 1962.

5

WILL THE NEW ELECTRONIC MEDIA REVOLUTIONIZE OUR DAILY LIVES?

JOHN P. ROBINSON

Being asked to visualize the future is a most attractive and agreeable task. Packing up all our present cares and woes, we can project our limited experiences into the world of twenty years from now and beyond. Moreover, these visions of the future are put forth under ideal conditions: Who will be around to embarrass us or force us to pay up if our forecasts turn out to be completely wrong?

That is the condition under which most forecasters work. Far more attention is given to the predictions of Jeanne Dixon and Jimmy the Greek or the market forecasters than to a systematic appraisal of their past performances. Seers are far more likely to remind us of the times they were right

The analyses described in this chapter were made possible by grants from the Policy Research and Analysis Division and the Measurement Methods and Data Research Program of the National Science Foundation.

than when they were wrong. As Ben Bagdikian put it in the broader historical context of *The Information Machines:*

> It has taken two hundred years of the Industrial Revolution for men to realize that they are not very good at predicting the consequences of their inventions: to the surprise of almost everyone, automobiles changed sex habits. Information devices are no exception: machines for mass communications produce unexpected changes in the relationship of the individual to his society.[1]

Ithiel de Sola Pool's recent collection of essays on the first hundred years of the telephone contains several lively examples of forecasts that proved to be off the mark. Like Bagdikian, Pool and his associates discovered that the bulk of the societal forecasts concerning the telephone was too conservative. But consider this radical prediction from an AT&T chief engineer in 1907: "Some day we will build up a world telephone system making necessary for all peoples the use of a common language, or common understanding of languages, which will join all the people of the earth into one brotherhood."[2]

Although one may dismiss his enthusiasm as company hype, there is a certain realism in another of his contemporary's forecasts: "Now that the telephone makes it possible for sounds to be canned the same as beef or milk, missionary sermons can be bottled and sent to the South Sea Islands, ready for the table instead of the missionary himself."[3] Given a certain allowance for hyperbole, this vision is now being realized by videocassette or cable television.

Pool's group documented a number of optimistic forecasts regarding the ability of the phone to reduce or eliminate crime. As awry as these forecasts now appear in hindsight, it was concluded they were less wrong than incomplete: "The impact on the amount of crime, however, depended primarily on what people wanted to do." This personal motivation is a theme on which I will elaborate shortly.

Pool and his associates also noted that few of these forecasters anticipated the powerful effects of the telephone on the overall structure of society. They argued that the telephone, along with the automobile, played a vital and unforeseen role in the development of the suburbs, a uniquely American use of space. At the same time that it promoted urban sprawl, however, the telephone also led to urban concentration. The modern skyscraper would have been an impossibility without the telephone.

Whereas the telephone and the automobile have transformed the spatial environment in which Americans live, it has been the nonwired communications technologies that have transformed the temporal aspects of American life—not where people spend their lives but how they live them from hour to hour and day to day. This social and technological revolution began about a quarter of a century ago, a developmental period roughly comparable to that of the telephone when the above predictions were made by AT&T people. After assessing where we now stand, the rest of this chapter is devoted to a few predictions about the future effects of this nonwired revolution.

THE TEMPORAL DIMENSION
OF SOCIAL CHANGE

Time is a fascinating and hidden index of the structure of our daily life and its quality. No matter how poor a person may be, he or she still has the same amount of time to spend each day as the rich person. The choices people make about how to spend their time therefore provide a solid behavioral indicator of what is of value to them.

For a number of years I have engaged in studies of what people do with their time. These studies include two national surveys of time-use patterns, one conducted in 1965 with 2021 respondents and one in 1975 with a different set of 2475 respondents. Anyone familiar with forecasting realizes that two points of time are hardly an adequate base from which to make predictions. But some of the trends in these data diverge so widely from the conventional wisdom about where this country is headed that they deserve wider dissemination and discussion. To broaden the perspective I also make use in this chapter of data from American studies conducted in the 1920s and 1930s and some time studies done in other countries.

The first impression conveyed by the studies is stability. Time spent on work, housework, sleep, and leisure by different groups of people all show considerable constancy over a period of time. This inelasticity is shown in Figure 1 which reveals how free time was used in 1965 and 1975 by the urban population and by basic subdivisions of that population—according to race, sex, employment, and marital status. It should be noted that this time period from 1965 to 1975 was one of considerable social change, which is hardly reflected in the data.

The second impression afforded by time-use data is the importance of communication in everyday life. Our time diaries show that people spend

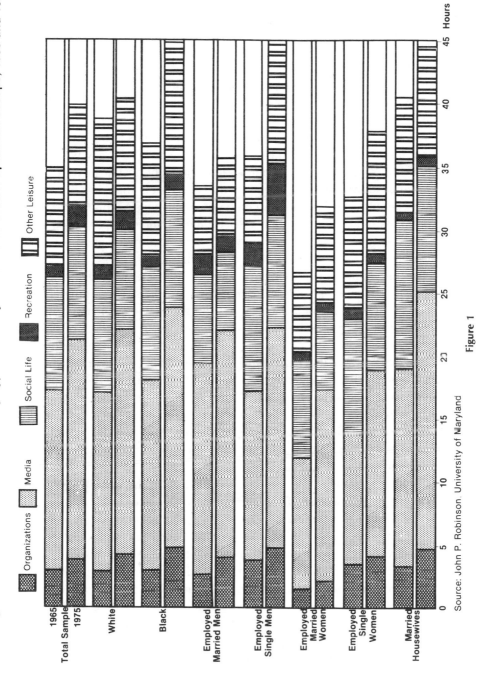

Average Hours Per Week of "Free Time" by Type of Activity For Selected Urban Population Groups, 1965 and 1975

Organizations · Media · Social Life · Recreation · Other Leisure

1965
Total Sample
1975
White
Black
Employed Married Men
Employed Single Men
Employed Married Women
Employed Single Women
Married Housewives

Hours: 0 5 10 15 20 25 30 35 40 45

Figure 1

Source: John P. Robinson, University of Maryland

63

an average of six hours a day in some form of communication. (It is also the case, of course, that people spend much of the rest of their waking hours preparing for or thinking about these activities.) Of that total time a little more than half is spent on interpersonal communications, with the other somewhat smaller portion being spent in contact with the mass media in one form or another.

It is this portion devoted to mass media that appears to be in the process of an important change. Extrapolating from the data, we may well be at a significant transition point in our society—namely an era in which more communication now takes place in mass channels than in conversation among people. New technology could potentially reverse this trend because it promises more user control over mass media and the greater likelihood of a two-way flow. At present, however, mass communication—with few exceptions—continues its one-way direction.

The third impression gleaned from our data is that technology in the home appears to have remarkably little impact on people's use of time. Though this may seem quite implausible, given the time-saving potentials of home technologies, it is nonetheless consistent with our first finding of stability. There seems to be little difference in the amount of time people devote to housework between those who have a dishwasher or clothes dryer and those who have not. This holds true even if we factor out of the data the greater ownership of technology by people who need it, such as working women or large families. It also holds true when we compare societies with low and high availability of technology; for example, Americans in comparison to people in Eastern Europe or Americans today in comparison to Americans fifty years ago.

There is, however, one significant exception in the data, as the reader undoubtedly has noticed: The largest shift in the use of time between 1965 and 1975 was for television, which reinforces the basis of my earlier remarks about television revolutionizing the temporal aspects of daily life.

There were three technological and marketing developments in the 1970s that could have been associated with television's increased usage: the spread and improvement of cable television, the introduction of video recorders (VCRs), and the distribution on a wide scale of color television.

Because 17 percent of our sample in 1975 were cable subscribers, we conducted an in-depth analysis of their daily lifestyles. We determined not only the amount of television cable subscribers watched but when they watched it and what other activities suffered as a result. To our great surprise the differences were minimal. Our cable subscribers did watch more television than nonsubscribers, but the difference amounted to *less than half-an-hour per week.* And this could well be explained by other factors that distinguished cable subscribers from nonsubscribers; for example, cable subscribers lived in areas in which greater viewing took place; they also had more free time. But, given a "television of abundance," people did not appear to watch more.

A possible explanation for our results may have been that many cable subscribers had only a limited number of channels or a high number of redundant channels. We therefore disaggregated cable owners in terms of the number of channels and the number of independent channels they received. Our conclusion was unchanged. In other words, even the presence of a diversity of independent programs did not seem to increase the amount of time cable subscribers spent with television. These conclusions are not unique. At least two other studies of which I am aware find either no difference or insignificant differences between cable subscribers and non-subscribers.

Though too few people in our 1975 study had video recorders, other studies indicate that the patterns generally confirm those that I have cited. The amount of time people with VCRs spend watching VCR tapes on their sets seems to be a negligible proportion (3 to 8 percent) of the total time spent viewing.[4] Also revealing is the finding that even among highly affluent people who can afford VCRs few are using them to skip commercials when they play back programs. Not many owners appear to use their recorders for long-run purposes, such as building libraries of treasured programs or movies. Most tapings are erased in less than a week's time. After reviewing the evidence from several studies of VCR owners, Agostino, Terry, and Johnson concluded:

> When the home medium of cartridge television was introduced in the early 1970s, it was heralded as "the greatest (media) revolution since the book" (Fortune), which would uproot "all political, educational and commercial establishments" (Newsweek). . . . The somewhat discouraging conclusion is that, so far, use of VCRs seems unimaginative, conditioned more by viewing habits learned in connection with traditional broadcasting and as yet not much affected by the great flexibility of the VCR as a home medium.[5]

There is certain encouragement to be found in this conclusion. We do not appear, counter to Staffan Linder's suggestion in *The Harried Leisure Class,* to be very apt slaves to our technology.[6]

These findings about cable and VCR are also consistent with the view that when people have the time they watch television—not television programs. Following Marshall McLuhan's general premise, content may be irrelevant in predicting people's viewing behavior. In this connection, however, it is important to note one exception. Color television did indeed seem to have an effect on increasing the time spent in front of the TV set. One reason, of course, is that color improved the quality of the image being received, regardless of content.

LONG-RANGE STRUCTURAL EFFECTS

Although I know of no formal studies of the behavioral impact of home computers or videodiscs, one can observe similar discrepancies between technological promise and user performance. What is particularly disturbing about microcomputers, however, is their highly uneven adoption by schools, which represent the developing ground for the next generation of home technology users. Computer literacy could well create another of the knowledge gaps that separate the information-rich from the information-poor.

At the 1979 conference on Working in the Twenty-First Century Walter Hahn discussed the implications of a paperless, cashless environment:

> Below the euphoria of the new technologies and global vistas are some grubby leftovers that must be dealt with. Who will do the dirty work? Compunications cannot clear streets, attend the ill in hospitals or slaughter cattle. . . . Will the even higher educational and intellectual demands of emerging technologies create a still larger functionally illiterate class?[7]

As in the history of the telephone, there is always the danger that we may be blind to the more subtle and long-range structural effects on society of new communications technologies. One possible outcome is discussed by other authors in this book. This is the sense of isolation and alienation that people may experience by the dispersal of work centers.

Our studies offer an interesting insight here. Between 1965 and 1975 television viewing increased and social interaction decreased *despite* the far greater enjoyment people claimed to derive from social interaction in contrast to media interaction.

Yet many possibilities are present, not all of them downbeat. In hindsight some communications researchers now feel that the important effect of television is to be found not in what media users think but in what they think about. The direct broadcasting of news to homes may also have rearranged expectations and power relations in society. No longer are the public's expectations shaped by the confines of their immediate environment.

The ability of the new video technologies to let viewers determine their own access and their own agendas may take a generation of experience to have any effect on reshaping societal consciousness of the information at our disposal. Even though today's cable viewers may not view more, the availability of more programs undoubtedly must lead to some improvement in the programs the public is able to see when they do view. Moreover, even though relatively few of today's VCR owners spend time watching prerecordings, the sales records of X-rated tapes make one wonder whether the VCR will challenge the automobile in terms of changing sex habits. We need to anticipate what new forms of consciousness will take place in the more mediated environment of the twenty-first century.

Notes

1. Ben Bagdikian, *The Information Machines,* Harper-Row, New York, 1971, p. 1.
2. Ithiel de Sola Pool, Ed., *The Social Impact of the Telephone,* MIT Press, Cambridge, 1977, p. 129.
3. *Ibid.,* p. 135.
4. Mark R. Levy, "Home Video Recorders: A User Survey," *Journal of Communication,* 23–27 (Autumn 1980).
5. Donald E. Agostino, Herbert A. Terry, and Rolland C. Johnson, "Home Video Recorders: Rights and Ratings," *Journal of Communication,* 28–35 (Autumn 1980).
6. Staffan Linder, *The Harried Leisure Class,* Columbia University Press, New York, 1970.
7. Walter Hahn, "The Post-Industrial Boom in Home Compunications," in C. Stewart Sheppard and Donald C. Carroll, Eds., *Working in the Twenty-First Century,* Wiley, New York, 1980, pp. 30–38.

III

INTERACTION:
THE HEART
OF THE MATTER

The interactive element in communication, a theme introduced in Section II of this book, is the focus here. All four contributors are in full agreement that anything worthy of being called communication must be a true interaction—a flow both ways. Otherwise, as one of the authors points out, it is merely a transaction in which information is passed along.

Beyond this there is little agreement, for "interaction" is a word with many subtle shades of meaning. The points of difference arise over how machines will relate to people and people to machines, not to mention the degree to which electronic communications will alter the nature of the human community itself and the ways in which people relate to one another.

Arno A. Penzias observes that up to the present technology has always demanded that people adjust to machines, but he believes this is changing. For the first time we have the remarkable capability of giving machines "friendly interfaces." Because we can now afford to use huge amounts of computer power for limited applications, it is possible to shift "the burden of understanding" from the person to the machine. Machines that respond even to vague or ambiguous voice instructions are being designed. And not far down the road are machines that will be able to talk to other machines as easily as we now talk among ourselves.

Anne W. Branscomb sees these electronic systems as giving people more effective ways in which to organize their lives by providing them with enormous amounts of vital information and making them highly mobile. She does not see people becoming more isolated by this dispersal but rather that communities will be restructured. The whole thrust of the new diversity provided by the media will be toward a greater choice among a wide variety of lifestyles and greater freedom to pick freely according to preference. The new kind of community, Branscomb believes, will be characterized by two things: less social stratification and greater equality of access to information even in remote locations.

Kathleen Nolan is dubious about the accommodation of machines to people, which is the underlying assumption of the two preceding authors. The history of technology shows, she believes, that people have a tendency to adapt to their own inventions and then to forget when and how they first came about. She is concerned that we will become more separate, confused, and angry with one another by misusing the new technologies instead of using them strictly for human ends—peace and freedom. She suggests that we turn for guidance to creative artists "who have the capacity for teaching old humans to become new humans, for teaching new humans to control their environment so that it does not control them."

Lawrence Halprin agrees with Nolan that communication in its real sense is holistic. We communicate not just through the intellect but also through the senses. "Everything you do as a whole person has to do with the meaning and the content of your message." The most creative experience occurs when people in a group are able to break through to a deeper level and develop shared ideas and intentions. The "common language of shared experience," says Halprin, is "a powerful force toward creating better communities and broad understanding among peoples."

6

FRIENDLY INTERFACES

ARNO A. PENZIAS

Three years ago the number of computers produced in the United States exceeded the number of people born in a year's time. Microcomputers on chips smaller than my fingernail already outnumber people in this country. It is not just the number of computers that is growing but their speed, efficiency, and power as well. This revolution in electronic intelligence provides us with the means to communicate with and control other technologies. It is now becoming possible to create friendly interfaces between people and machines.

It seems to me that machines will play a decisive role in our future. In the past societies like that of the ancient Greeks were able to develop a culture that still awes us today. The Greeks did this without automobiles, power plants, or computers, but they did have one thing that we lack: a plentiful supply of human slaves. Technology today has made it possible for us to exploit machines rather than people.

Unfortunately, many of our present machines appear to complicate our lives and intimidate us. Instead, machines should give us more control over our own lives, help us to deal with a complex and bureaucratic world, and free us to do the things that only people can do. Machines should begin to accommodate people instead of people accommodating machines.

It is much easier to tell a human being what to do than to instruct a machine. Human beings understand speech, can sense movement and body language, and, most important, can ask questions and think for themselves. In the same way the friendly interfaces of the future will make it easier to tell machines what to do. They will understand speech, sense users' needs, and ask for more information when necessary to clarify instructions or obtain more information. Although they will not be truly intelligent, such friendly interfaces will act in an apparently flexible and intelligent manner. They will allow us to communicate with other machines as if the machines were friendly, attentive human beings.

We are rapidly progressing toward friendly interfaces in a number of technologies, both at Bell Laboratories and at many other research and development organizations. But most of our present technology still demands that people adapt to the needs of machines; for instance, we sit in front of a keyboard with fingers poised to strike the keys and quantify our instructions by physically moving controls. Most people do not want to bother with manual skills such as typing, but they still need to interact with machines.

Broadcast technologies such as radio and television also demand that we adapt to machines. We synchronize our lives to conform with broadcast schedules—we finish dinner in time for the movie. These technologies, however, do not allow two-way communication. Perhaps that is why live theater still flourishes. Actors are willing to accept lower salaries, and people are willing to pay high ticket prices for live theater. They want feedback—laughter, applause, even coughing or shuffling of feet. Surely people will expect friendly interfaces to provide interactive capabilities.

Another example of a technology that requires human adaptation is the printed page. As a communications medium the printed page is quite limited. It cannot reproduce the author's intended tone of voice, body movements, or facial expressions. And we must laboriously convert thoughts to printed characters. Many attempts have been made to transcend these limitations. Written dialogue is meant to reproduce the sound of speech. Glossaries, appendices, footnotes, and references are meant to anticipate the readers' questions. Today electronic analogs of these devices make it possible to transfer information—whether text, graphics, sound, or video—to approximate more closely human-to-human interactions.

Technologists dream of a human-machine interface that will offer flexibility, feedback, and freedom from physical contact. Such an interface would let us talk to the machine and have it understand and respond. The machine would obey even vague or ambiguous instructions; it would select among alternatives, it would make decisions based on experiences, and it would seek additional direction when necessary. The burden of understanding would shift from the person to the machine.

Designing and building such an interface will not be easy, but there are grounds for optimism. The cost of computing power has been declining at an astonishing rate. The computer I used as a graduate student filled a three-story building. Today I can buy an equivalent amount of computing power for a few hundred dollars and I can hold it in my hand. As recently as four years ago the most powerful minicomputer, medium-sized and in the $100,000 class, had sixteen-bit words. That means that it could recognize numbers only up to a certain size; therefore its efficiency was limited. More advanced thirty-two-bit computers were very large and cost about $1 million. This year Bell Laboratories has under development a thirty-two-bit computer on a chip as big as a cornflake. The little machines are catching up to the big ones.

The result is that we can now afford to use huge amounts of computing power for relatively small applications. We can substitute dedication for intelligence. We can design a computer for a specific task instead of for general-purpose use. This "dedicated use" approach lets us use so much computer power that the machine's limitations are effectively hidden. Thus the fact that the computer lacks true intelligence is merely an engineering obstacle, not a fundamental flaw.

However, overcoming this engineering obstacle is a formidable task. The human brain is a uniquely powerful instrument. Computers must be stretched to the limits of their capabilities to replicate functions, such as speech recognition, that nature accomplishes with subtle simplicity. We now know enough about the sounds, the patterns, and the structure of speech to instruct a computer to recognize some spoken words; for example, we have trained a computer to recognize enough words spoken by a particular user to confirm airline reservations.

To perform a more difficult speech recognition task, like converting words into text at dictation speed, would require as much computing power as 1000 of today's large general-purpose business computers working together in perfect harmony. Fortunately, we don't need 1000 large general-purpose computers if we're willing to trade flexibility for power. We can create a specialized computer optimized for a specific task.

Until recently the cost of creating such specialized computers was prohibitive for all but the most important applications. By the use of advanced computer-aided design techniques, however, it should soon be possible for almost anyone with a particular need to design a specialized computer to meet that need. Computer users will act as computer designers, using one kind of friendly interface to help create others.

At the same time our ability to let computers talk to one another is growing rapidly; for instance, in the last few years tremendous progress has been made in turning glass into wire. A glass fiber cable as big around as my finger can carry at least 50,000 digitally encoded telephone conversations simultaneously in both directions, and it can do so much more economically than copper wire. In the next few years a network of glass fiber cables will become a superhighway for information that will connect hundreds of intelligent switching machines which the Bell System already has in place to handle ever-increasing voice and data traffic. Thus machines will be able to talk to other machines as easily as people now talk to other people.

In the Bell System we already have computers talking to other computers all the time; for example, suppose an engineer wants to design a new central switching office. First he types into a computer terminal a slightly more complicated version of the question, "What do I need to build this switching office?" Then that computer asks another computer, "What is the climate in the new location, and what size air conditioner will the new office need?" That computer asks still another computer, "What size air conditioner is in stock, and how soon can it be shipped?" The machines talk among themselves to accomplish the routine tasks that make telephone service a little more efficient and less expensive. But it is not easy. Designing the software—the complex set of instructions—that allows these dumb machines to understand others is, in fact, very difficult. If there is one small error, all the computers just fold their arms and glare at one another. So the interfaces between machines also need to be made friendlier.

Friendly interfaces between machines could make our lives easier in all kinds of ways: for instance, a computer in my car could talk to the Pontiac dealer. When it senses that some part is in trouble, it could notify the garage to make sure that the right replacement part is ordered before I take the car in for repair.

By sensing users' needs and communicating them to the appropriate mechanism, friendly interfaces could help to provide people with services similar to those provided by a secretary, butler, or handyman. They can analyze clues derived from various sensors, communicate with faraway data bases, and look for similarities in previous situations in order to make decisions. Although they will not be able to make value judgments, they will seem to be intelligent.

How well we succeed in creating these friendly interfaces depends on our use of human intelligence. We must continue to seek new knowledge, exercise our ingenuity, and encourage curiosity. Scientists and technologists must recognize and consider the wants and needs of society, and society must choose its technological options wisely. A friendly interface between people and machines requires a friendly interface between a free society and its scientific and technical community.

7

FINDING ONE'S PLACE
IN A MULTIMEDIA SOCIETY

ANNE W. BRANSCOMB

Communications and community are derived from the same Latin root, *communis*, meaning "common," and are closely related. To communicate, according to Webster, is "to share, impart, or partake," whereas community means "a body of individuals organized into a unit manifesting some unifying trait." Thus many sociologists and philosophers look on communications as the *web of society*—the unifying force. When God spoke to Adam and Eve, the communications environment was small, and the communications were simple and direct, face to face, in the same language. Even so, Adam and Eve did not share the same understanding of the utility of eating fruit.

Today television can communicate instantaneously and simultaneously to a global audience. Some 538 million viewers shared the view of the first lunar boot stepping on the moon. That their reactions might be similar or their values shared is less likely. They came to the experience with vastly different social, economic, and political histories.

I take my text from a question asked me by a college student who lived with us as a baby-sitter when my children were small. Donna found that she could not cope with our round table. At every meal each of us sat at a different place. She had come from a rigid and authoritarian family. Her father, an Army colonel, presided over a table at which each child had an assigned place. After several weeks of rotating at random around our table, she came to me in great dismay and inquired, "Mrs. Branscomb, where is my place? I must know which is my place."

This was not a problem in the Middle Ages because people then were born to a certain status, a certain position, and a certain place from which it was difficult to extricate themselves. What they could wear was dictated by that status: A prosperous merchant could dress no better than a knight with half his annual income. Today the cultural cues are wildly cacophonous. We can choose among many religions, lifestyles, work environments, and spouses. Nobody knows which is the right place or which is the most comfortable place or where to find a place to adapt to the changing political, social, and economic environment.

Extrapolation tells us that the world of the twenty-first century is going to be even more diverse, more complex, more pluralistic, and more libertarian. Thus conflicting social and moral values will confront us with a veritable smorgasbord of philosophies. How shall we cope with this carousel of choices? Like Donna, we may ask ourselves how do we find a place?

SIMPLICITY VERSUS COMPLEXITY
Managing the Information Explosion

The communications environment of the tribe or village was small, closed, and simple. The computerized information society will be just the opposite—complex and open—but it cannot be any more perplexing than the present babel with which we are surrounded in the print media.

I can remember when a letter in my mailbox made my day and when the arrival of the Sears, Roebuck catalogue was an occasion for great joy. Not anymore. Today I find myself drowning in a sea of direct mail. I find myself traveling more and more because returning to my permanent mailing address is so painful. If I stay away longer than a week, I can hardly get through the obstacle course of catalogues, bills, and solicitations that obstructs my ingress. In this sea there is a wealth of information, some of which I want, some of which I do not. It is difficult to tell which is which, but I admit to being a curious person. Recently I threw away all the catalogues that had accumulated over a period of several months except those that really captivated my interest. There remained some sixty-eight in my shopping bag for future perusal.

There are some who decry the computerized society of the future because it will invade their privacy and record their every move. Only urban academics who have never lived in a small town could believe that this is some new threat. My own experience causes me to assume that a fairly accurate profile already exists of my income, travel, consumer preferences, hobbies, and favorite charities. It is clear from my magazines and mail that I am a lawyer interested in communications, foreign affairs, housing, and estate planning; that I am a gardener, housekeeper, birdwatcher, sailor, skier; and that I belong to a rural electric cooperative. The direct mailers have not decided yet what I won't buy, so they send me everything. The political types either cannot figure me out or do not give up easily: I receive political mail from a complete spectrum of organizations from the far left to the far right. When I purchased a CB radio several years ago, my mailbox was filled with solicitations from CB clubs and magazines. After a visit to the computer show, my mailbox filled up with solicitations from Radio Shack, Apple Computer, Personal Computing, and so on.

At one point I became so distressed over the overstuffed mailbox phenomenon and its companion problems of paper pollution and wasted time that I went to the extreme of asking the direct mail association to remove my name from all lists. (The association refuses to remove you selectively, which is clever, because few people want complete isolation; they just want control over the subjects they like.) My ploy did not work. Within six months our mail had quadrupled. All the direct mailers who hadn't known about us just added us to the list.

There is one more step I might take. The Supreme Court has decreed that my home is my castle. If I really want peace and quiet, I can stamp all that mail "obscene." But I am reluctant to do that. Besides, it is often educational to browse through even the most unlikely catalogue, for that's how I discovered I could buy a backyard satellite receiver for a mere $3595. And how would I have known that I wanted one otherwise?

What I need is not more information but better information management. Computers in the twenty-first century will help me to accomplish this. I look forward to a day of electronic mail when I can scan the titles of the catalogues and solicitations as they roll across my CRT. Then I'll push the delete button and they'll all disappear into the ether whence they came. Now that may be an economic disaster for the garbage collector, but it will save a lot of timber and free me for more creative work than slicing open envelopes. Computers will help me find what I am looking for at the time I need it without spending hours shuffling through a lot of papers.

The information utilities that will provide access at the time and place of my choice are well on the way to realization. Tymnet offers data users several dozen data services in twenty-eight countries and forty-three states, according to its catalogues; Cybernet offers hundreds of sophisticated software programs to twenty-four nations; and the Telenet network maintains more than 100 commercial service bureaus and computerized information services. The best known data provider is Lockheed Systems' Dialog, which describes itself as an interactive information retrieval service that makes available more than 100 data bases, primarily abstracts. The Dow Jones News Service claims more than 8500 computer subscribers whom they serve with financial data, stock quotations from *The Wall Street Journal, Barron's,* and articles from *The New York Times,* all for $50 per month and $40 per hour connection time. *The New York Times* Information Bank provides more than sixty publications at a time charge of $120 per hour. Several legal data bases are available to subscribers such as Mead Data's *Lexis,* West Publishing's *Westlaw,* the Air Force's *Flite,* and the Department of Justice's *Juris.* Doctors can tap into *Medline,* a bibliographical medical service provided by the National Institutes of Health. A public data network for individual users called *Source* purports to be an information utility that provides access to biorhythms, menus, electronic games, wine lists, UPI news services, financial analyses, and, among other things, a personal mailbox.

One would quickly suspect that this confirms the theory of information poverty in an economy of information wealth. The information utility will be no different from electric or gas utilities. Consumers will pay for the information they consume, and those with more specialized needs will hire specialists to sort out the information they need. *Reader's Digest* is second in circulation only to *TV Guide,* which helps us sort out the television programs we want to see. We already have digests of digests in the print media; abstracts of professional journals are a thriving business, and big corporations circulate to their executives a special compilation of news articles related to their institutional interests.

Information vendors will supply whatever market is large enough to aggregate itself into a profitable sales segment. The most remarkable aggregation of markets in the history of information products has been the worldwide aggregation of movie consumers, which permits producers to spend as much as $20 million on one production. Market aggregation has marked the development of the data-processing industry from the beginning when time-sharing services became available to small users. Just as we seldom, if ever, find a movie or a book designed for an audience of one, we are unlikely to find an information utility designed for one. Yet a microcomputer for less than the price of a car may provide enough data-processing capability to organize one's personal life. In trying to find ourselves in the twenty-first century we will have at our disposal technological systems capable of reaching millions in our global electronic society or we can retreat to the data confined on our own sets of floppy discs. This is not unlike the technological range that exists between the encyclopedia and the personal diary.

SOCIALIZATION VERSUS ISOLATION
Expanding One's Family

Several years ago cartoonists liked to project the evolution of *homo sapiens* by picturing a large blob sitting in a chair with earphones attached to very large ears, bulbous eyes glued to the television screen, and arms and legs reduced to tiny appendages. The idea, of course, was that the new technologies would produce physical isolation.

Let us remind ourselves that there was nothing more isolated than pioneer life on the prairie, where the nearest neighbor might be a day's horseback ride away. One has only to read *Centennial* or *The Thorn Birds* to understand how important social interaction was to those who lived in such circumstances. Betty Friedan has documented the intellectual isolation of the suburbs where one is surrounded by what my daughter chooses to call "rug rats." I have a college classmate who combined her own imagination with that of a sympathetic husband to tackle this problem of intellectual isolation. They were both journalists, but she was a journalist-turned-housewife, whereas he, as a science writer, covered diverse professional meetings of scientific specialists. Discovering that these scientists never shared their predictions of the future, husband and wife started a cottage industry at home by publishing a modest newsletter in which they assembled the papers from the professional societies that the husband, Ed Cornish, had covered. The World Future Society has grown from that modest birth fifteen years ago to an organization of more than 50,000 people. In 1980, in cooperation with the Canadian World Future Society, it held its third and largest international conference in Toronto. More than 5000 attended from 45 countries, and some 1000 speakers held forth at more than 400 sessions.

The Cornishes are not unlike the pioneer family on the prairie: Both husband and wife pitched in to provide for their family's survival. But the modern version produces a different level of intellectual stimulation, opportunities for travel, and interaction over long distances with their professional colleagues who include some of the most provocative minds of our day: Alvin Toffler, Isaac Asimov, Herman Kahn, and Hazel Henderson. The nascent World Future Society may be in primitive form the kind of extended family or network of sustaining relationships that will prevail in the twenty-first century. They are information providers, pioneers of the information age.

A hundred years ago the patterns of interaction were more often dominated by church or ethnic background, and usually by both. In a community analysis I conducted in Middletown, Connecticut, in anticipation of a new cable television system, we found that ethnic and religious patterns from earlier migrations still prevailed and centered markedly on churches. There were several Roman Catholic churches in the community. Only one, however, offered masses in Spanish, whereas two offered Italian, which was spoken only by an older, established minority. In suburban New York my house hunting quickly sorted out the residential patterns (now somewhat muddled but still apparent) of entire satellite cities populated historically by one ethnic group or another—the Irish Catholics in Rye, the Jews in Harrison, the Quakers in Chappaqua, the blacks in New Rochelle, the Italians in Valhalla, the Poles in Ossining, and the WASPS in Greenwich. The trend today, however, is generally toward identification with professions or sports activities rather than with religious or ethnic groups. It is said that on the East Coast within fifteen minutes after meeting a New Yorker you will know whether that person is into commodities, law, medicine, politics, or banking. On the West Coast, by contrast, Californians are identified as golfers, skiers, sailors, or surfers.

The trend toward affinity affiliation rather than geographical bonding of communities was recognized by the Lynds in *Middletown*, their monumental sociological analysis of a conglomerate midwest city in or about 1929. Already the advent of the automobile and telephone had transformed the social interaction from the "dropping in" variety which characterized a local community limited in social interaction to the ambit of foot or horse and buggy. In contrast to the social life of the nineteenth century, which had revolved around church socials and neighborhood taffy pulls, the social life of the twentieth century was, according to the Lynds, increasingly sophisticated:

> Today, lawyers, doctors, bankers and ministers, and now and then other professional groups meet in their several associations to listen to papers on the details of their work. . . . The new civic clubs (e.g., Kiwanis and Rotary) . . . with golf, bridge, and motoring are the leisure interests toward which business men gravitate. . . .[1]

Before widespread literacy and the mass distribution of newspapers the church and the pulpit constituted the major communications system. Today, thanks to magazines and the support of television advertising and sports coverage, there is a proliferation of sports-oriented groups, whereas professional journals and conferences cement professional and business affinities. Intense interest in these pursuits is focused by the physical interaction of major media events (e.g., the Olympics or the Nobel Prize).

Apparently there is little evidence so far of social isolation imposed by the new communications environment. Rather the opposite seems to apply. There are more opportunities to interact physically as well as electronically. Indeed, the conversations that led to this chapter were conducted via trans-Pacific satellite but culminated face-to-face on a warm, sunny day on the beach at Waikiki.

CENTRIFUGAL VERSUS CENTRIPETAL FORCES
Access to the Global Marketplace

As we move from the technology of scarcity in communications to the technology of abundance it might be prudent to be mindful of the centripetal forces that encourage greater diversity in lifestyles, in both the print and audio media, and to anticipate a similar diversity of choice in video messages. There is nothing so binding as the shared message of the Bible, Torah, or Koran delivered from the pulpit or Bimah or Minbar to an audience that cannot read. Where in the proliferating technologies of communication, with their almost infinite variety of information messages, are the ties that bind? What kinds of communities can we expect to emerge from a global pluralism?

As its Christmas special last year Neiman Marcus advertised an earth satellite receiver for $23,000. When I met Mr. Marcus at the opening of the store in White Plains, New York, he admitted that only one had been sold. The Marcus seal of approval is an implicit recognition of a trend downward from the more than $100,000 that the original saucers cost. This year an innocent sportswear catalogue offered clothing and equipment needs for sports enthusiasts from running bras to trampolines. There, hidden next to the binoculars, was item number 450, which recommended that "TV's best programs are relayed by satellite—enjoy them all with an earth link from Channel One." The price was down to $13,000 for a five-meter dish and $11,900 for a dish of about three and one-half meters.

If your backyard saucer has a clear view of the southern sky, free of obstructions, your saucer would bring in some two dozen different satellite services, which include four superstations (WTBS in Atlanta, WGN in Chicago, KTVU in Oakland, and WOR in New York), three religious services (Trinity, PTL, and the Christian Broadcasting Network), a Spanish-language service (Galavision), two children's services (Nickelodeon and Calliope), the U.S. House of Representatives, two special sports services (Thursday night baseball and Madison Square Garden), several movie channels, a community service network, Ted Turner's Twenty-Four-Hour-Cable-News network, and, of course, Home Box Office and Showtime. Which of these services were being delivered to paying customers and which were free was not mentioned. This small detail will not go unnoticed by the movie producers and sports program packagers who are screaming "piracy" at every alleged infringement of their intellectual property rights.

The fact is that direct broadcasting from satellites is becoming a reality even if unintended. More proof came from my Christmas visit to Colorado. First I discovered that the electronics store was selling, like hot cakes, a series of articles describing how to build your own earth station for less than $1000. Then my hot-tub salesman came in with a surprise announcement that he was adding satellite saucers to his electronic boutique. He wondered if I didn't surely think Santa Claus should deliver one to my hillside site for a price no more than the latest Detroit X car. Now American Value, Inc., of Rolling Meadows, Illinois, is offering a ten-foot dish for $3595, which folds up like an umbrella. I suppose you take it along on your camper.

Comsat—the people who brought us the geostationary satellite in orbit for an international common carrier service—is now prepared to deliver programs to your rooftop; three different services are promised, twenty-four hours a day. Channel A will offer Superstar, a movie, and general family entertainment (the network challenge). The Channel B "spectrum" will be a cultural affairs and children's channel (PBS competition). Channel C promises "viewers' choice" sports events and special programming such as "Dollar Sense," "Soapbox," "Lecture Hall," and "Government at Work" (a challenge to pay cable). Thus the pattern set by Sir Arthur Clarke, who foretold the coming of the geostationary satellite in 1945, nears completion. He sits in his favorite haven in Sri Lanka where, remote and isolated, he speaks to the world through his books; and the world beats a path to his door. However, because he has not yet acquired the "comsole" (the communications console) that he has predicted, there will be no sequel to *2001: A Space Odyssey*. Sir Arthur refuses to leave Sri Lanka for more than two weeks a year and Stanley Kubrick refuses to fly in an airplane. Thus these two lifestyles are incompatible, and the world is minus this productive partnership. But we are reminded that, given a few more technological breakthroughs, the collaboration could probably take place satisfactorily in the twenty-first century without interrupting the lifestyles of Kubrick or Clarke.

Not many of us can claim the talent or success of the Kubricks or Clarkes of the world, but each of us has a little ambition for professional success not entirely squelched by the moralities of today. Some of us also have a nostalgic longing to live in rural but not intellectual isolation. And many of us are beginning to realize our dreams. Our house in Colorado is in El Jebel, which means "the mountain" in Arabic. El Jebel is in the middle of one of those growing disaggregated urban settlements that can only be described as "exurbopolises." Ours is called Down Valley. It stretches from Aspen to Glenwood Springs, a distance of some forty miles, and includes many small townships such as Old Snowmass, Snowmass Village, Woody Creek, Basalt, Emma, and Carbondale. It has no city center, no mayor, and no city council. Indeed, three counties have portions of their boundaries in Down

Valley—Pitkin, Eagle, and Garfield. But it has a life, a soul, and loyal and enthusiastic inhabitants—it even has growing pains like the great urban centers.

Unlike the people in suburban areas of New York or Middletown who cluster around their church and ethnic group, Down Valley society is held together by sun and snow and skiers and hikers. Those who love the mountains and have opted out of the more structured environments of city life are adopting a laid-back lifestyle. Some would say they are "copping out," but that is not entirely accurate. Among my neighbors and friends are an accountant who gave up his profession for carpentry and cross-country skiing and an educator who became a ski instructor. Others include a professor of chemistry, a vice president of a large oil company, and a stockbroker. Some of them commute regularly to New York, Los Angeles, or Washington, others only occasionally. Some stay put in Colorado. But they all claim Down Valley as their chosen home. Nor is Down Valley substantially different from Cape Cod, which caters to Washington journalists and MIT professors. The only difference is that Cape Cod attracts sailors rather than skiers. These are only two of the geographically disaggregated communities that are developing not only in the United States but around the world. Queensland, Australia, is building a cultural community along the Great Barrier Reef; the Cotswolds, around Oxford University in England, are not unlike Delmarva (the Eastern Shores of Delaware, Maryland, and Virginia) and Central Florida. These "exurbopolises" are made possible by a combination of high-speed telecommunications and jet travel and are characterized by two features: less social stratification and greater equality of access to information in remote locations than has ever been provided in all of history.

Teddy Roosevelt called the Presidency "a bully pulpit," and his cousin Franklin turned radio into the bully microphone by reaching out directly to his constituents all over the country. Across the Atlantic others who saw the radio as a device for forging national unity or spreading international socialism almost overwhelmed the western world. Thus we encounter the good and the bad news. The new technologies are able to bind and at the same time offer a far greater variety of informational opportunities than Hamlet ever dreamed of. Some are quite appalled by the advent of the electronic church. The most appalled are local ministers who are still operating on the "drop in" theory of the ministry. Some abhor the commercialism of religion. Yet the electronic ministers are merely religious entrepreneurs who are taking advantage of the new technology of communication. The television screen is the ultimate bully pulpit for preacher or president alike. What more competitive marketplace of ideas could one wish for than a television dial that can be switched at will from Anglican to Zen with all the variations in between? If there is an ultimate truth, this may be a far easier way to find it than in pilgrimages to the Vatican, the Wailing Wall, or Mecca. In a multimedia society we may be confused by all the proselytizers, but at least we have freedom of choice—if we preserve the political system that permits us to choose and an economic system that can afford the price of competition.

DEHUMANIZATION VERSUS HUMANIZATION
Unleashing The Human Mind

The cry of the student protests of the 1960s was "Don't fold, bend, or mutilate," the words found on the IBM punchcard. This represented a genuine fear that individuality, personal lifestyle, and freedom of expression were being lost in a society that was becoming more and more mechanized. Indeed, entire nations have reacted similarly. In France the Nora Minc report and in Canada the Clyne Commission decried the loss of jobs by computerization. The Swedes and Brazilians are disconcerted about importing technologies that will change their lifestyles.

As much as we sympathize with workers who fear obsolescence of their skills, we must nevertheless recognize that there is nothing more dehumanizing than assembly-line production in which the same laborer puts the same nut on the same bolt for forty years. It is in the changing organization and specialization of job responsibilities that we unleash people to engage in more intellectual and humanistic endeavors. The information economy in which we now find ourselves is one in which the primary activities are services based on education, entertainment, and technology. Ours is a society that values knowledge more than manual dexterity. An example of the way in which computers have become a humanizing influence is offered by computer-aided manufacturing. This does not necessarily mean robotics. At many points in the manufacturing process computers are accelerating the design process, managing the material inventory, and personalizing the particular product. We can virtually design our own automobiles, albeit from a half-dozen General Motors shapes but with infinite variety of color, trim, and accessories—a far cry from Ford's standardized black Model T. Moreover, in the not-too-distant future by the use of holography machine tools will be programmed to give you shoes that not only fit perfectly but are as personalized as those made by a twelfth-century cobbler, maybe more so.[2]

In the classroom the computer can provide one-to-one teaching without a hostile or disapproving class listening to one's recitation. This may be a much more humane way to acquire some skills, though obviously not the best way to learn those that require socialization. Learning can be made more fun. Take a look at the electronic toy market. Texas Instruments has come out with a number of teaching toys such as *Speak and Spell* and the *Little Genius.* The Marconi International Fellowship will be awarded this year to MIT professor Seymour Papert who has developed a special computer language for children called LOGO. The major advantage is that the child teaches the dumb machine rather than being subjected to the authoritarian atmosphere in which a learned adult is teaching dumb children.

More important than the role of machines in personalizing education and consumer products is the freeing of humans from the boredom of inhuman tasks. I leave for the economists the problem of allocating equitably the fruits of the labors of the robots, but I welcome a society in which more and more human beings can share the exhilarating effects of creative endeavors. I can only echo Isaac Asimov who has said, "The twenty-first century may be the great age of creativity in which machines will do the commonplace work of humanity, and human beings will be free at last to do things that only human beings can do—to create."[3] If we solve the allocation problem, then we may anticipate a flowering of the arts in the twenty-first century that will put ancient Greece to shame.

The danger is not, as I see it, from Big Brother in 2024 but from ourselves. Computers and robotics will be able to free us from onerous tasks, but I fear we may become lazy and let our brains atrophy. We should never lose sight of the fact that a human brain is vastly more complex and capable than a computer. We have billions of brain cells we never use. Our brains have capabilities that are still unmatched and unlikely to be matched by computers.

Professor Frank Kochen of the University of Michigan recently spoke on "The Role of Computers and Information Economies." I was in hearty agreement with much that he said, but not with his remark that if only we could translate all of our problems into higher computer languages the computer would find solutions for us.[4] I was offended, for whatever a computer can do and do well it cannot find solutions to our problems. Only we can do that with brain power that is unique in its capability for insight, innovation, inspiration, and integration.

Magazines, clubs, and professional societies can expand our human networks. Satellites and television can give us access to a global information marketplace. Computers can help us manage complex systems and free us from dehumanizing manual work. But only we can find ourselves.

Notes

1. Robert S. and Helen M. Lynd, *Middletown—A Study in Contemporary American Culture,* Harcourt, New York, 1929, p. 301.
2. Michael L. Dertouzos, "Individualized Automation" in Michael L. Dertouzos and Joel Moss, Eds., *The Computer Age: A Twenty-Year View,* MIT Press, Cambridge, 1979, p. 39.
3. Isaac Asimov, "The Permanent Dark Age: Can We Avoid It?" in C. Stewart Sheppard and Donald C. Carroll, Eds., *Working in the Twenty-First Century,* Wiley, New York, 1980, p. 10.
4. Frank Kochen, "The Role of Computers and Information Economies" in *Proceedings of the Pacific Telecommunications Conference,* January 1981.

8

THE SUBVERSIVE SUBSPECIES

KATHLEEN NOLAN

To paraphrase an actor and writer of another century, it sometimes seems that there is much ado about the new technology with a lot of sound and fury all signifying information overload. I will not add to the overload by giving my view of the impact of multichannel, optical fiber, and satellite transmission—and matters of *that* kind. I want to talk about something very different—about the impact of the "new human" on the technology of the twenty-first century.

Certainly humans are developing as fast as the new technology. Or are we? Have we stopped evolving? Nobody has been having conferences about that, so I thought I had better bring up the subject. This immediately raises a question: What is the new human?

But before we get into this there is another question to consider: What is the old human? Are we not called the human animal? Certainly we have all noticed that we do act like animals from time to time. So what separates "human" from "animal"? In the 1930s the American poet and philosopher Gertrude Stein came up with the answer in a piece she wrote for the mass media of the day: "All animals have the same emotions and the same ways as men," she said in the *Saturday Evening Post*. "Anybody who has lots of animals knows that. But the thing no animal can do is count."

So that's it. We are not animals because we can count. That brings up still another problem. If counting separates us from animals, then are we machines? Machines can count. Most of them can count better than I can—in fact *all* of them can. What makes us different from our machines? My answer is this: The thing that differentiates humans from machines is play. Many machines may have the same logic as humans. Anybody who has lots of machines can see that. But the thing no machine can do is create and the thing no machine can know is play.

If humans are creatures who can count and play, the next question is, what impact will this odd species have on the latest new technology that they have invented? Will they play with it and break it? Or play with it, get bored, throw it away, and go back to playing with one another? Or will they quit playing? Some of us are worried that humans will stop playing and start spending all their time with machines. Some of us are worried that they could *become* their machines.

Now this is something to worry about because the other thing we know about humans is that when we want to survive we are very adaptable. Many years ago, when our evolutionary relatives kicked some of us out of the trees, we started adapting to our new environment on the ground in order to stay alive, and we created several new technologies that no animal had ever created before. We created a new way of walking—on two feet; we created the fashion industry—starting small with just animal skins as our first line; we created indoor-outdoor heating by rubbing two sticks together; we created interior decorating and design by fixing up the walls of our caves. It is my theory that the most oppressed hominid in those days, the one ostracized from the trees, was the most artistic.

The most stunning and revolutionary creation of these old-time artists was the new communications technology. It was called language. Now we did not write down how we did it—writing was a later, new technology—but here is what I imagine happened.

Some of us were just playing together or playing and counting together. Then we started making meaningless sounds and combining those mean-ingless sounds in the same way over and over again until some of the combinations of sounds started to mean something. We created words, and it was no longer merely a call system that we communicated with. We created a new technology that allowed us to say an infinite number of things by just playing and imitating the repetition. That was the beginning of the word, and creating the word was fun.

Pretty soon everybody was spending most of his or her time using the new technology. Not long after, everyone had a new vocal apparatus to make better sounds, whereupon some humans invented the words "be quiet" because so many humans were talking all the time. Others who simply could not stand it went off into the wilderness to meditate. Next all the humans went their different ways, and they kept talking and kept changing the way they talked. It then became difficult for some groups of humans to understand other groups of humans, and everyone knows the history from there. Because they did not understand each other, they did not like each other. Some of them thought they were superior to the others—the males thought they were superior to the females, one color thought they were superior to the others—so they started fighting. And whenever they started fighting it was always the most creative among them that said, "Oh, come on, let's play."

We are still inventing new technologies even though we have not yet figured out the impact of that old technology, language. We do not know exactly how we did it. And this is a reason why technology is something to worry about with humans. They adapt to their own inventions and they forget how they started adapting to them. Will we do with the new communication technologies what we did with language? Will we become more separate, confused, and angry with one another? Or will we use the new means of communicating for human values, for global peace, for individual freedom?

I have given this question a lot of thought and I know what I want. I want human beings with a new kind of consciousness to have an impact on our new technology, to control it and use it, and to teach other humans to control it and use it. I want us to use the technologies we invent not to confuse and injure one another but to nurture ourselves and support our highest ideals. I want humans to create some new habits: the habit of creative consciousness, the habit of freedom. Just as sheep are in the habit of being sheep, so humans—the most adaptable of all species—can learn to get in the habit of being creative.

We have got to know who can help us do that, and they happen to be from a peculiar subspecies of human that has actually been around for a long time. Some people would even call this subspecies aliens, perhaps subversives. At times throughout history the members of this subspecies have been ostracized and ridiculed. But they have the capacity for teaching old humans to become new humans, for teaching new humans to control their environment to prevent it from controlling them, for teaching humans to choose rather than be chosen, for teaching humans to feel and to think, to survive, and to be alive. This subspecies is the creative artist. If we do not want to become more and more homogeneous, to have less and less content in our lives, to become a people devoid of emotion, then we must allow the creative subspecies to teach us.

You have to listen very carefully to understand how creativity works. You have to pay attention. First you have to dig deep down inside you. The artist, the performing artist, the musician, the painter, the sculptor, the architect, the author—all must touch some reality deep inside in order to create. It may be quite different from anything you have ever experienced, but you have to trust it—you have to trust your own machine. It can be painful at times because when you touch that truth it certainly is scary.

It is often said that artists are born, and that is true. Artists *are* born. All people are born with creative minds and instincts, and we can be trained to use our instruments. People do exercises to sensitize themselves, to hear better, to listen—to really listen to what somebody else is saying, to learn to know when they are being manipulated, to learn to manipulate. I propose that we change the system—infiltrate it, if you will—so that we can let the artist in and the artist in every one of us out.

I believe that the most creative among us should be in control of the new technology. Two things are needed if we are to have a choice about what we will be experiencing in the twenty-first century. The high priests of technology must be concerned not only with money but with the content of the message, the human ideals expressed. As receivers of the message we must develop a creative consciousness with which to tell the difference.

The question is really one of choosing the colors of your lives. There is gray and there is vanilla and chocolate chip and pistachio. There is nothing wrong with a little gray and a little vanilla, but you have to mix it up once in awhile. That is what the artist is capable of doing: mixing the colors, making the choices. There is no one in this media-conscious, information-overloaded society who has not heard a lot about the process of self-realization. The popular psychology books, the magazines, and the whole new-age consciousness industry from est to the corner spiritualist sell it. But the artist has never joined that marketplace. I propose that we do.

When people say they have no choice, that is not true. But choice means mixing colors, taking risks. It means seeing and facing change. The most significant difference between the creative and noncreative human being is the ability to change, to be mercurial, to be a chameleon, to be whatever one chooses to be, to turn oneself inside out.

Mere survival in little boxes is not necessary when the choices are there. We can learn not to be faceless audiences but to be participants, to be communicators. We are not yet at a crossroads and we can go either way, though perhaps time is running out. A survey a few years ago asked children if they had to choose between their fathers and television what would they give up. They chose to give up their fathers. That's scary.

It is at these dangerous times in history, when the species is in trouble, that we need the artist the most. The artist is the least prejudiced because the artist is looking at all the colors and seeing the value in all that is around us. Let us quit being afraid of the artist. Let us admit that in the educational system, in science, the most exciting work has always been done by people who have dared to deviate from the straight and narrow line.

The formulas of our mass media do not encourage the artist to play. If the audience does not change, we will continue to have only those established formulas in which an explosion occurs every seven minutes because we have not tried to touch anything deep but have only wanted a surface experience. Life has no art except in the experience of those who respond to it. When we eliminate the possibility of that experience, we eliminate art itself.

Those who are in control of the media say that people want to remain numb and unaware. That is debatable, but if that is what the people want then we need to teach people that they do not have to accept and want so little. If the machine that we want to function better than any other machine is the human machine, then we have to oil it, take care of it, and nurture it. We must let it have senses and smells and roundness and feeling and touching—not just one-dimensional experiences. That is what we're in danger of losing.

I was once in a class at the Neighborhood Playhouse in New York with the greatest acting teacher in the world, Sanford Meisner. I was right off the *Goldenrod Showboat* and had been acting all my life, so I was not really a smarty kid who thought I knew what I was doing. Meisner put me up in front of the class and said, "What's on your shoe?" And I said, "Nothing, sir. There's nothing on my shoe." And he said, "What's on your shoe?" And I said, "Nothing, there's nothing on my shoe." I just kept looking him straight in the face and saying that there was nothing on my shoe. Before he finished with me I was down on the floor with my shoe off, turning it upside down, inside out, feeling it, touching it, sensing it, holding it, smelling it. And he said, "Now do you know?" And I said, "There is nothing on my shoe." And he said, "But you didn't know that before because you weren't examining fully the possibility that there was something on your shoe."

Walter Brennan—I did *The Real McCoys* with him for seven years—was playing old men when he was in his twenties. Where did that ability come from? It was his ability to touch that old soul within him. We all need to be able to touch the old soul. More important, we need to touch the innocence within us. What is happening is that more and more we are having our innocence removed.

The school my son goes to is using the art form to teach history, to teach the relationship between mathematics and music. What they are teaching is creative consciousness. This school is a little place in Bath, Maine, with 200 students, and they have been struggling for years to define what it is they are doing. Why is it that all these kids are singing and dancing and doing sports better than they ever have before and are faring better than their contemporaries at all the very structured New England schools? It is because they are learning by the experience of doing, they are using all their senses and not just becoming corporate robots. How do they learn about the French Canadians? They are singing the songs of the French Canadians, going to the workplaces up there, talking to them, sharing their history, and then dancing with them. The school has never been able to define why it works. I know why it works. They are teaching creative consciousness; they are teaching the students to see all the colors.

So we need to listen to the artist. We need to let the artists become educators. We need to let our new technologies become art forms. Just as poets, orators, and dramatists have made language an art form, so must artists lead the way in making the new media not simply preprogrammed mass computers of mediocre entertainment but new art forms that express human values in ways appropriate to the particular medium of communication. The corporate media business moguls need to let the artists play.

Artists are the proletariat of this communications revolution. We are the subversive subspecies. Artists are in touch with the fact that humans need to keep in touch—truly in touch with one another and with the magical reality all around us.

At another time in history, when another upheaval in our society was causing many thinkers to worry and wonder what was to become of us,

Abraham Lincoln helped inspire and guide us through those troubled times. He wrote: "What's good for the working people is good for the nation." With the immense upheaval and change happening in the communications field, I want to amend his statement with my own, very serious prediction: What's good for the artist is good for the nation.

The person who has perhaps contributed more to our understanding of our relationship to technology than anyone, the late Marshall McLuhan, has said, "I don't want people to believe what I say. I just want them to think." So do I. So does my peculiar subspecies, the artist. Is that too subversive? Is that too much to ask?

9

THE REAL MEANING
OF COMMUNICATION

LAWRENCE HALPRIN

I understand that a one-way dialogue across a network of mechanical devices, or even person-to-person, is not any form of communication at all. As far as I am concerned it is a transfer of information that is auxiliary to the act of communication.

I did not start out thinking that way. When I was younger I believed that if I had a message and was sure in my soul of the morality, ethics, and rightness of what I was going to say then everybody would buy it. I soon found that this was true among most men I talked to, but the women did not accept it at all. I have learned that the hard way from the women in my life: my wife, then my daughters, now my granddaughters. They all tell me the same thing. "Husband" or "Daddy" or "Granddaddy," they say, "get off this *telling* us about things and giving us information; tell us how you *feel* about something. Tell us what your experience is. Then we'll understand what you are trying to say, and we can feed back to you and deal with you. Otherwise, you are laying a trip on us."

This is not merely off-the-wall California talk. It is a human response that I find wherever I go. Two-way communication becomes even more important when there are groups of people because it deals with a whole network of things other than information. It deals with emotions, with the past, and with the present. It deals with symbols, myths, and rituals, with biases, and with the acculturation of viewpoints of many kinds. It has little to do with information but a great deal to do with how you touch people in a profound sense and how they touch you.

I suspect that a failure in two-way communication is the cause of the national trauma we experience in our dealings with foreign countries. They keep misunderstanding our view of what we stand for because we do not really communicate with them. We give them information, we give them goods, and we tell them what we think they ought to do. But we really have not connected with them emotionally. There are ways in which this connection can be made. A lot of time and effort has gone into discovering the processes and techniques that can be applied consciously in opening up two-way communication between people, and that is what this chapter is about.

Take the issue of evolution in California right now. The theory of evolution is conveyed by lecturing, which is a form of one-way information transfer. What everybody tends to forget in this kind of lecture is that beyond content are implications, highly emotional ones, in the message about evolutionary theory. For many people it simply violates basic attitudes, and *telling* them that we are related to the apes cannot induce them to feel that in fact we are. The emotional and religious trauma is too great. Here is a learning process that attempts to induce us as children to accept what somebody is telling us. Whether we accept it or not depends a good deal on the culture from which we come.

We are instructed about right and wrong, good and bad, and because we are taught these things on an informational level we have no opportunity to decide for ourselves. This is a perfectly valid engineering approach. The engineer explains to people that a freeway is the best thing for them because it is going to move cars from one place to another in the fastest possible way. Never mind what it does to your community. The instruction tends to backlash. Because people have not been involved in making the decision, they revolt.

The old-fashioned linear process is one to which we are all accustomed. In planning or in personal relationships most of us have a linear approach by which we accept all input as long as it agrees with our point of view. We have a predetermined goal. Of course, one of the major problems is that it builds up frustrations in people because they feel that nobody is listening to them, and rightly so. This comes up over and over again, especially with young people. It can be overcome by the process called "active listening," which is active in the sense that you listen carefully and then you feed back to the other person what you have heard. Here is a good example: "I know that you believe you understand what you think I said, but I am not sure that you realize that what you heard is not what I meant." If you ever try active listening, you will find that it is wonderful for children.

I was brought into it when my little daughter came over and kicked me in the shins and said "I hate you." One way to handle this is to haul off and spank, which is what many people do. That drives the frustration deeper. The other way is active listening; you say to the child, "I understand that you're really angry at me." The child says, "Yes. I'm angry because you paid more attention to Johnny than you did to me." Then you say, "Well, I understand that you are angry because you really want me to pay attention to you." This goes on and on until the child finally understands and says, "Daddy, I really love you." That is a form of active listening. I recommend it to you because it gets through tremendous frustration.

Listeners will often hear the content but will not hear the feeling behind the content of the message, and it is the feeling that is often more important. This holistic approach to communication deals with more than intellect. What we usually do is a matter of the intellect, but that leaves out a lot of important messages from the senses. There is sight and there is smell; there is the kinesthetic sense and the sense of touch, all as important to us as pure intellect—very often more important. With the holistic approach everything you do as a whole person has to do with the meaning and the content of your message.

In the planning work in which I am involved I have created a technique of communication called the RSVP cycle. Very briefly, in this acronym R stands for resources, S for score, which has to do with how you generate activity and participation, V for value actions and feedback, and P for performance. In this process of cyclical involvement with people you can generate creativity, which is what really counts. Without a sense of participation there cannot be any group creativity. Both the creativity and the participation are the result of what can be described as an ecological concept that forms the core of RSVP.

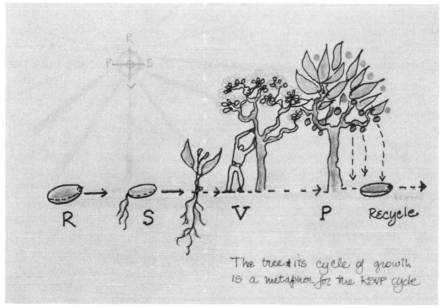

The tree & its cycle of growth is a metaphor for the RSVP cycle

Figure 1

Figure 1 is a simplified explanation of the concept: the R, or resources, is thought of as the seed of a fruit; score is taken from musical terminology and has to do with the development of the theme, or the growth of the seed; the value action is the human intervention and decision-making part of the process, when you prune the tree to make the fruit more edible; and finally, there is a recycling when the fruit falls to the ground and grows into a new tree. This places emphasis on revealing sensitivities, on awareness, and on communication with yourself as well as with others.

Whenever communications are discussed there is always a lot of talk about a common language. If you can put people in positions in which they can share experience—experience on a deep level—that in itself is the most powerful form of common language that I can possibly imagine. Those who work in foreign countries, even though they do not know the language well, find after a while that they hardly need language to communicate, provided that they experience things together. Kinesthetics are important because of the universality of movement; no less important are painting, design, poetry, and all forms of expression. These use not merely the intellect but all the senses, those vital ways of communicating with others as well as with yourself. Open conflict is also an important vehicle for communication. If we hide conflict, we tend to suppress a vital source of human creativity. We tend to make the mistake of ignoring conflict by not giving it formal acceptance as part of our experience with others. When conflict can be expressed and given a position—an important one is our communication with one another—it becomes a creative device. Often I find in workshops that it is by conflict and impasses that we can break through to do innovative work. You will notice the body language of the person who is being attacked. The result of this intense and emotional conflict is a breakthrough in creativity.

I was recently asked to teach the RSVP cycle in Japan. One member of the group was a psychiatrist and psychologist at a university in Tokyo. At the beginning he was very straightlaced in his intellectual attitude and communicated with the folks in the workshop in a manner that fitted a formal portrait he had of himself. Gradually he eased off because he realized that there were ways of communicating other than talking. After movement sessions and many other activities his self-portrait was much looser, more integrated, warmer, more related.

You can achieve the same results in other ways. You can produce self-portraits by drawing what people's neighborhoods are and what they would like them to be. Dealing with what your neighborhood is and what your environment does to you is a form of saying some important things about your life that none of us has ever been able to articulate. As a result of these self-portraits people often begin to break through and make major advances not only in themselves but in how they relate to the world.

I can cite an actual example that occurred in the town of Charlottesville, Virginia, during a "taking part" workshop a few years ago. The people in the community went out and participated in awareness walking. Often people say to me, "I have lived in a community for years, and I've never really seen it until we dealt with it on this multifaceted level that you've spoken of." After taking their walk these people came in and made drawings of what they thought ought to happen, thus communicating with one another and coming to an understanding of the various alternatives. This kind of sharing of ideas and intentions with others is an important part of holistic communication.

Within a year the business district was redesigned as the result of the workshop—a communication process transferred into reality. The resulting plaza-like setting came about because of intense communication among people making their own selections about what they wanted. You can do this not only with physical things but also with intellectual ideas, processes, theater pieces, or whatever. This form of communication and participation encourages all kinds of creative things to happen.

Fort Worth, Texas, offers another example of this process at work. There had been years of impossible decisions and of frustration about what to do in this community when it was finally decided to make a breakthrough. I proposed that we deal with it at an RSVP workshop of leading citizens. They went out and saw places that had been beautiful when they were children but were no longer because of the technical requirements of the U.S. Army Corps of Engineers. They were not all together at the beginning; they didn't all agree at the end; but despite diverse points of view they learned how to arrive at a consensus through this common experience.

These people did not understand the transit problem because they were accustomed to going from their air-conditioned country clubs to their air-conditioned cars; therefore what was the point of public transit? I asked them as part of the "score" phase to walk across town for two hours during the day at lunch time and try to get from one place to another on foot or by existing mass transit. When they came back there was full agreement that they were going to put in a new mass transportation system because they had experienced 110 degrees in the shade—and trying to get around in that heat was debilitating. This is the kind of common experience that communicates itself to people.

The basic human needs that were expressed by this kind of group through personal participation come up over and over again. I could go on about the myths that are enacted, the archetypes that are revealed, the shapes and forms of things that occur when people deal with one another in this communicating process and release the intense creativity that is within them. Instead I will end with a simple statement. If we really communicate with one another, nothing is impossible.

IV

STRATEGIES FOR

MANAGEMENT

What will happen to organizational structures as sophisticated electronic information systems are put into place? What kind of steps should management take to install these systems? How long will it take and what are the problems that will be encountered? These are some of the questions explored by the contributors to this section of the book. Collectively, they have little doubt that a radical change in organizations is on the way because of the revolution in electronics, and they have confidence that the results—on the whole—will be highly beneficial both for institutions and for people. Their concern is with the strategies that managers must adopt to facilitate the inevitable.

F. G. Rodgers believes that information systems give greater impetus than ever to the evolution of the horizontal organization, as opposed to the traditional vertical or hierarchical type. In seeking out expertise, management will deal increasingly with people on the basis of what they know rather than where they are in the corporate hierarchy.

The vitally important data that make for business success or failure are generated today by some thirty million principals, including scientists, engineers, accountants, financial people, and other professionals. Rodgers observes tremendous opportunities to increase efficiency and productivity in the handling of this information by the assimilation of these people into an organization based on electronic systems. But this is inhibited by inertia in the marketplace caused in part by the fact that 70 percent of management's time at present is spent maintaining older systems.

Gail E. Bergsven says that the challenge lies not only in introducing the changes but also in prevailing on management to revise its modes and attitudes. For one thing the conventional geography of work will change. Traditional workplaces will not disappear, but much of the work will be shifted to "alternate" worksites, which include satellite locations and even workers' homes. For some workers this will create problems of isolation and loneliness and will require a new kind of effort in understanding and caring on the part of managers. "Remote" work will also cause many managers to feel threatened because of the loss of visible control. "The role of the manager," concludes Bergsven, "will no longer be that of sitting at the apex, but rather that of strategic planning and consensus-gathering, both of which put a premium on communicating effectively."

Louis H. Mertes observes that many professionals who welcome the mechanization of clerical staffs have deep emotional problems accepting the mechanization of their own work. The key, he says, is to identify the "early adapters" who enjoy moving out ahead of others and to organize them into small pilot projects. Because the investment in these projects is relatively small, it is "unnecessary to force success or postpone the natural death of a failure." The psychological barriers can be bypassed by encouraging each professional to develop his or her own style of participation. The ultimate payoff for electronic systems is that they release individuals from having to be in the same place at the same time to get their work done; each can proceed at his own pace at a convenient location often remote from others.

Peter G. W. Keen believes that the new electronic systems will bring about a radical alteration in the structure and processes of most organizations well before the end of this century. Information is now seen as a corporate resource analogous to that provided by the corporate financial staff. Telecommunications also allows us to see organizations more abstractly as systems for information processing. Keen agrees with Rodgers that these systems provide a powerful push toward the horizontal rather than the vertical corporate model. Electronic networks permit centralization-with-decentralization, shift the influence of central staffs, eliminate spatial restraints on organizations, and redistribute data. "That," concludes Keen, "adds up to a radical set of forces for change."

10

THE CONSTANTS OF CHANGE

F. G. RODGERS

We live in a time of paradox, contradiction, turbulence, opportunity, and, above all, change. If, in the thirty-one years that I have been with International Business Machines, there is any one thing that I have learned it is that change is always with us and that it also has a profound psychological impact. To the fearful change is threatening because things seem to be getting worse. To the hopeful change is encouraging because things are going to be better. To people who believe in themselves the effect is still different: Change acts as a stimulus. There is a conviction that one person can make a difference in this world or in a given corporation or institution. People who look on life that way might be called "difference makers," and it is these people whom we are all looking for; we are eager to attract them to our organizations.

This chapter is addressed to these difference makers, the people who will somehow help to guide our various enterprises through the maze of change that we call progress. I see nine specific changes or challenges that we face now and that we will face in any future that we can conjure up. Not all of these trends are directly related to communications, but taken together they form the background against which we will be establishing communications policy for our organizations and for the nation. These challenges will affect us whether we are makers or users of communications equipment and other electronic systems or whether we are acting in our individual capacities or as representatives of some institutional entity.

If we expect our communications programs to work, we will have to bear in mind at all times the trends that I single out for attention. My list has been developed from a long intimacy with the rapidly changing world of computers, which are among the chief instigators of change in our time, and from travels that have taken me continuously around the globe for a number of years.

CHANGING VALUE SYSTEMS

No matter where you go in the industrialized or developing worlds it is impossible to avoid the realization that value systems are undergoing enormous change. Everywhere people talk about the quality of life. You cannot pick up a newspaper or magazine without seeing phrases like "more is less" and "less is more." A few years ago it was fashionable to challenge the work ethic, and I happened to think at the time that it was very much overdone. What is happening today, I believe, represents a fundamental change in basic attitudes toward work and its meaning in our lives. Regardless of their job levels, people are trying to seek the right balance in their lives. They are saying, yes, a fair day's work for a fair day's pay. But in the process we will not sacrifice our families. Nor will we pass up involvement in activities outside work that give us a sense of social responsibility. This is happening to all of us.

Another major change is also occurring. People no longer want to be told *how*, and that includes my own children and IBM's customers. They want to be told *why*. They demand some rationale for actions; they want to know why things are going on. They want to be able to go home at night and say, "What I did today was worthwhile."

That is a very positive thing. I used to do a lot of lecturing on college campuses in the 1960s, when this was not too popular, and I see a big difference between then and now. In the 1960s I found students who had the ability to understand what the problem was but were unwilling, for whatever reason, to articulate a solution. Today people are just as unhappy with the ethics and conduct they see around them, but at least they are willing to offer constructive suggestions.

This probing attitude actually has nothing to do with a person's age. I used to think that a preoccupation with the quality of life was something that had to do only with the young people corporations were hiring from the college campuses. I now know that sensitivity to changing values is not a prerogative of the young. People of all ages share it.

EDUCATION AND MANAGEMENT STYLE

The second phenomenon of our times is the rising tide of education. Some people think of this in terms of ten million students on our college campuses, but I do not mean that. The IBM Corporation or any institution, regardless of size, is typified primarily by vertical structure. How does an organization really function? Organizations operate increasingly on a horizontal basis. This means that you deal with people on the basis of what they know rather than where they sit in the hierarchical structure of a business.

The futurists like to describe this as the world of the knowledge seekers. If you have a problem, you try to solve it by going where the best knowledge and the greatest expertise exist. You do not have to climb through the layers of the organization to get there. Automated systems help you to tap into data bases, thus making that information easily available.

Long gone are the days of making fundamental business decisions using only data of a historical nature. What people want is simply the ability to tap into any aspect of their businesses at any point, identify problems, and—one hopes—solve them before they become disasters. A good information system should also be able to reveal the basic strengths of a business and how they relate to the areas of weakness.

We are finally reaching a stage at which computer systems, regardless of where they come from and from what company, will greatly amplify man's intelligence. They will not necessarily replace it. Management must take an active role in the information systems that will bring this enormous force to bear on the situations and problems that need to be resolved. Management must be involved in the decision-making process that determines the goals of these systems; this should not be left up to the technical people.

THE EFFECT OF ECONOMIC ILLITERACY

The third force behind change is economic illiteracy. We are getting a fresh breeze out of Washington now, but it is not enough. I visit some of the finest, most prestigious universities in the United States, including the business schools, and I ask the students a fundamental question: Out of every dollar you collect in gross income, what do you pay in taxes? Or I ask the reverse question: Out of every gross dollar you make in income what are you taking out in profits? The answers are discouraging.

Seventy percent of the American public today say that corporations large and small charge excessive prices for their products. Twenty percent, according to a recent survey, say price fixing is a standard mode of operation.

Another recent survey asked 1000 students from our finest campuses to rate professions according to their degree of ethics and morality. College professors came out number one, perhaps rightfully so. Doctors and engineers should feel fairly secure because they were in the 62 to 63 percent range; lawyers rated 40 percent. Business people were at 19 percent. That was about 15 percent above advertising practitioners, who were rated about the same as labor union leaders.

There is a fundamental misunderstanding here. A tremendous amount of work needs to be done to improve economic literacy in this country. It has somehow been overlooked in spite of the general improvements in our education level.

GOVERNMENT REGULATION

We constantly hear that government and business in the United States do not work together as they do in Japan but sometimes foolishly draw apart. People ask what regulation really costs businesses today. The Business Roundtable tried to determine this point by looking at forty-eight companies regulated by six agencies of the Federal Government. They found that the incremental cost of regulation here totaled $2.6 billion a year.

In the pharmaceutical industry two-thirds of the research and development outlays are eaten up by compliance with federal regulations. Here are some more discouraging statistics: In 1974 the budgets of forty-one regulatory agencies in the Federal Government amounted to $2.2 billion, and at the end of 1978 those budgets had risen to $4.5 billion, a 19.5 percent compound gross rate. This was substantially beyond the gross rate of revenue increases of the companies these agencies regulate.

No one in business wants to abolish the OSHA laws or other important regulatory codes. What industry is saying is that some balance between regulation and economic growth is essential.

THE ANTITRUST LAWS

There is a certain ambivalence in our society in regard to the antitrust laws. Many people feel antitrust has been put in place to protect competitors from failure; I think the antitrust laws are there to foster free enterprise, and that means there have to be some winners and losers. The trouble is that nobody wants to be on the losing side. College professors all want tenure, labor union leaders don't want layoffs, business people want nothing but profits.

I subscribe to the interpretation of the antitrust laws that says that a corporation can reach any level of penetration of the market it it does so by excellence. Then you let supply and demand and good old-fashioned marketing prevail. In addition, I learned a long time ago that you cannot strengthen the weak by weakening the strong. You cannot help the wage earner by tearing down the wage payers. You cannot build character and integrity in people by taking away their initiative. You cannot help somebody permanently by doing something that they could or should be doing for themselves. That is the essence of free enterprise.

TECHNOLOGY AND THE MARKETPLACE

I have lived through the vacuum tube, the transistor, solid logic technology, and now the wonderful world of microminiaturization. A new type of low-temperature or superconducting circuit that operates at thirteen picoseconds and switches information at seven picoseconds was announced recently at our research laboratory. You cannot go any faster than that, although some scientists say we're going to find a way to circumvent that problem. We are only at the midpoint of where technology will take us. We will be talking not about storing 64,000 circuits on a chip but about 100 million circuits on a one-inch ceramic chip by bringing temperatures down to minus 273 degrees centigrade.

We're now seeing work being done on signature verification and handwriting analysis. Perhaps the area of greatest capability will be voice recognition. By the end of this decade there should be machines able to operate on a 5000-word vocabulary in normal language, not special dialects. Then there is electrostatic printing and ink jet technology. Instead of 200 characters per second, as at present, we will be operating at 500 and 1500 characters per second—in color.

We are also seeing advances in communications centered on satellites. Fourteen commercial satellites are now aloft and at least eighty-six more will be up by the end of this decade. Terrestrial communications costs are coming down every year by 11 percent, and those of satellite communications are dropping at an annual rate of 40 percent. Computer technology is improving its price performance by about 25 percent a year, thus giving people more for less.

That brings us to a potential Achilles heel of the electronics industry. Is there enough capability out there in the customers' offices to absorb these new technologies? All known studies show that something like ten major office applications are waiting to be systematized by data processing and communications, but 70 percent of the customers' programming staffs are spending time maintaining older systems and simply cannot get to that list of potential uses.

Another significant area is that of human factors, or ergonomics. In Germany, for example, you cannot sell a display terminal that does not have what is called a "flicker-free" capability and a keyboard that is separate from the CRT itself. There will be many decisions made in the marketplace in the next three years on the basis of human factors and ease of use. What Arno Penzias calls "friendly interfaces" will be as important in selling equipment as price and performance are today. People in the electronics industry recognize this significant fact, and many have insisted that this thinking become ingrained in the development function. A look at the new twenty-four-hour transaction terminals in banking will show that some very exciting things are coming into play.

CONSUMER ADVOCACY

One of the great forces of change is consumer advocacy, and I contend that there are ways of making it work with you rather than against you if you are in business. This is illustrated by the story of the universal bar code, the optical character-sensing device used in supermarkets.

It has been only in the last year or so that it has taken hold. In the beginning comparative shoppers would go into a supermarket and look at the shelf price; then they would look at the bar code, and because they could not read it they thought the price was there but intentionally disguised. Of course, the price is not there but in the computer. This is a better way of handling price changes than to have someone running around trying to mark all the cans and packages. By education and training, industry is now convincing people that the computer or any other device is nothing more than a tool. That idea has at last begun to take hold. The moral is that you cannot underestimate the individual in this world of advanced communications and advanced technology.

PRIVACY AND SECURITY

Another challenge that business must wrestle with is the issue of privacy and security presented by electronic communications. How much information do people want to give to others? An example of one of the things we have tried to do is to recognize that perhaps we can't solve all the problems of privacy but that we have the responsibility for demonstrating some sort of action and leadership. In 1974 we published an ad entitled "IBM Principles of Privacy" in *Time, Newsweek, Forbes,* and a number of other publications.

Subsequently, we developed a rather extensive employee privacy program that we implemented within IBM and communicated to our employees. One of the things we did was to stop the use of social security numbers for internal identification. We also opened employee personnel files to our people. A lot of things can be done to ensure privacy by the proper handling of employee information.

Security, as contrasted with privacy, presents very different problems. There are ways through encryption and algorithmic techniques to make sure that only the right person has access to particular information in electronic communications. A major force for change, however, may be electronic funds transfer systems and use of all the other things people envision at some point in time are going to occur in the area of electronic communications in the home and office.

PRODUCTIVITY IN THE OFFICE

The area in which electronic systems will have the greatest impact over the next five to ten years will be the office. Just a generation ago only one in ten workers in the United States had white-collar jobs. Now about one of every two workers are white-collar, and by 1983 the ratio will be about six or seven workers in every ten. This puts an enormous importance on productivity in the office. Specifically, the rate of office productivity growth has been under 4 percent, whereas productivity in the factory over the last ten years has been at a 91 percent rate. One reason for this is that the office is undercapitalized. The farmer works with $53,000 in capital goods per person, and only 3 percent of our population produces all of our food and a significant amount that is sold abroad. Each factory worker has about $35,000 in capital goods. What we have given to the office worker is only $2500 in capital goods. A study by the Stanford Research Institute indicates that by 1985 there will be $10,000 in capital expenditures for every administrative worker in America. That adds up to a fantastic opportunity for everybody.

Another tremendous opportunity is presented by the thirteen million secretaries and administrative workers who will be in the labor force in the mid-1980s; there are obvious advantages in mechanizing their work. But the great area of opportunity is occupied by the thirty million principals, twenty-two million of whom are involved in authoring material. I am thinking of engineers, scientists, accountants, professionals—all of us who deal continuously with the creation, handling, and review of data and information.

A few years ago it was estimated that if we could improve productivity in the United States by 1 percent we could generate some $600 billion in new gross national product. That is tough to do, but it is entirely possible with technology in the hands of creative and intelligent people. Productivity is said by some to mean working harder, longer, and under increased stress.

C. Jackson Grayson, head of the American Productivity Center, says that productivity is simply output over input. Peter Drucker says that it is the first test of management. Here is still another definition of productivity, one that I happen to like: the untapped potential generated by the creative interchange of new ideas, new technologies, and intelligent people who are skillfully and consistently managed.

THERE ARE ALSO THINGS THAT DO NOT CHANGE

Go back to the turn of the century, which really was not too long ago, and look at the twenty-five top companies in business at that time. Then look at those same twenty-five companies to see where they are today. If you do this, you may be startled to find that only *one* is still in that select circle, and it merged with a half-dozen or so other companies. Several of the twenty-five are not even in business any more; the remainder are still functioning but nowhere near their old positions of preeminence.

Historians, scholars, and all the rest of us try to analyze what makes the difference between success and failure. Why does one organization survive and flourish when in the same line of work another fails? When an organization succeeds, people say it had a charismatic leader, that it put more into R&D, that it was sensitive to what people happened to want and need. All that can be true. As for the companies that failed, it is said that they turned inward or lost their perspective or were overwhelmed by marketing myopia. All that can be true, too, yet there is something more.

The real difference between success and failure is how well a business organization or any institution succeeds in developing the talents and energies of its people, regardless of the positions those people hold. An organization needs a set of principles and beliefs to guide people, and these principles need to be articulated and clearly understood within the business. There are some basic principles that IBM regards as fundamental in creating an environment in which people do their utmost to devote their energies in the right direction.

First comes respect for the individual. That is simply making sure that just as the company respects the individuals whom it employs so the employees have respect for the individuals with whom they deal. Respect for the individual is shown by the way switchboard operators are trained to handle phone calls or how receptionists are trained to greet people.

Another vitally important premise is that the organization must insist that it give the best service of any company in the world. Unfortunately, when we get good service in the United States we are surprised; it ought to be the other way around. Finally, an organization should expect excellence from its people.

If you set principles and beliefs in place, if you measure people and reward success and penalize failure, then you will help your organization to weather cyclical periods or the test of time. This can be achieved only if the organization is willing to give the respect that all individuals deserve. There are two things in a business that you should increase out of proportion to its growth rate: One is education and the other is communication—telling people what is really happening. They deserve no less.

11

HUMAN RESOURCE PLANNING IN A NEW AGE

GAIL E. BERGSVEN

Some people say nostalgia isn't what it used to be. Given the global uncertainties we face today, it may be even more appropriate to suggest that futurism isn't what it's going to be — especially when we're talking about the subject of human resource planning in a new age. Predictions are especially precarious in this transition period between the industrial and information ages because the most infinitely complex variable is also the most important: human behavior.

During the early years of computer development those who looked ahead quickly recognized the challenge to human adaptability. At a time when there were only thirty operating computer systems in the nation the man who coined the term "automation," John Diebold, said:

> The real potential, and the enormous problem automation poses to the manager, is that the environment in which the enterprise exists is changing, rapidly and completely. As the goals, aspirations, needs, and wants of the individual shift, and shift again and again through the human social change induced by automation, the economic realities that sustain the enterprise will change. In

other words, the great meaning of automation to the manager is to be found
in the *social change* induced by automation.

Ironically, one of the first changes was that Diebold's term "automation"
lost favor after being linked with the threat of jobs being lost when new
technologies were applied. Today, several decades into the new age, we
are made aware of daily individual and managerial challenges in the application of computer-based technologies in ways that accommodate human
as well as practical economic favors.

Whether we are also more sensitive remains open to question. A report
last year by Working Women, a national clerical association, arrived at this
conclusion:

> Unless clericals organize to influence office automation in the 1980s—while
> the technology is still in its formative stages—the health, well-being, quality
> of worklife, and employment of women workers will be sacrificed for the sake
> of "corporate progress." Imposing the conditions of the assembly line on this
> country's 20 million clerical workers would be criminal.

As you reflect on that statement consider that the National Office Products
Association has predicted that the sale of office machines will grow from
$17.6 billion in 1977 to $37.5 billion in 1983 and to $83 billion by the end
of the decade. The office alone will present a major challenge to human
resource planning in the new age. The challenge will not be just in the
manner in which an organization introduces change but in how well management can adapt its own modes and attitudes.

A survey of business managers done recently for *Newsweek* revealed that
63 percent identified word processing as automatic or memory typing and
not, as expected, as centralized processing of office information and communication. Another study pointed out that many executives fear that the
advent of the electronic office means that they will have to give up personal
secretaries—and lose status. Still another study revealed that managers often
resist using the keyboards of access devices that provide them with essential
information.

All of these findings suggest the depth and complexity of old assumptions
and habits that must give way to new and necessary technologies. The
challenge is considerable, especially if we are to apply effectively the new
tool becoming available to solve the deepening problems of productivity.

In the future offices and worksites as we know them will become impractical and obsolete. Data-processing and communications technologies will combine to bring about a greater emphasis on the organization and application of information rather than the cumbersome mechanics of production and dissemination. This means that the geography of work will change.

I agree with futurist Alvin Toffler when, in his book, *The Third Wave,* he suggests: "The most striking change in the third wave civilization will probably be the shift of work from both office and factory into the home." As the work product changes, as employees seek new labor markets, and as more people by necessity enter the work force traditional workplaces will inevitably change—perhaps not disappear but gradually disperse to satellite locations and in-home sites, to "electronic cottages." Work will go to the workers. And as the workplace changes, managers must adapt

THE CHANGING WORKSITE

Control Data is among the companies already experimenting with the alternative worksites of home and satellite offices. The Professional Services Division—composed of engineers, systems analysts, programmers, management consultants, and educators—is in its second year of an alternate work station program. What we have learned from it so far will serve as a foundation for adapting to the larger worksite shifts to come.

There are three primary reasons to institute an alternative worksite program: first, the energy saved by substantially reducing commuting; second, the forecast of a continued, even worsening, period of escalating prices and static wages; third, the continuing decline of communication costs, which will more than offset the initial expense required to create a network of electronic cottages.

Control Data had still another reason to begin such a program: disabled employees. Like any large corporation, the company is faced with the enormous cost of benefits for homebound workers. Two years ago Control Data started a program to retrain disabled employees to do computer-based work at home, thereby reducing costs while providing productivity for people who might otherwise not be able to work. The success of this experiment,

Figure 1.   John Schatzlein, Administrator/Counselor for the Homework program, evaluates and selects potential Homework participants for external contracts; he then provides counseling and support for the participants.

called Homework, has now led the company to examine the larger issues of alternate worksites and to start a more far-reaching program in the Professional Services Division.

THE MANAGER OF THE FUTURE

As for the manager of the future, one trait is mentioned immediately and repeatedly by executives working on programs: flexibility. The work force emerging today is highly diverse in terms of age, sex, race, education, background, and ability. Among other things, the work force is getting older, and minority workers are increasingly filling entry-level positions. A manager of the future must be a good communicator if he or she is going to reach and motivate the changing work force; he or she must be oriented toward people rather than solely toward production. Thinking will be holistic rather

Figure 2. Robert Peters, a technical writer at Control Data, trains Homework participants at home so that they are able to reenter their traditional worksite after appropriate adaptations have been made.

than linear, both in regard to projects and to people. Management, if it is not already, must become a profession rather than a step or a reward.

In the new information age our business is the gathering and distribution of knowledge, a uniquely human product. Employees become a critical and uncertain link in the process, and their management becomes crucial. As a result, the traditional backgrounds of top corporate officers—marketing, finance, and law—will soon be joined by a fourth: human resources management.

The people we manage are no longer solely motivated by dollar amounts. Nor is their loyalty given solely to their companies; it often attaches to their professions. People are more likely to think of a career not as long-term employment with one company but as a progression of challenges and advancements in a particular field and, if need be, in several companies. The cost of training replacements is expensive, both in time and productivity.

Initially, the form of work led to problems in productivity and quality control. How does one measure and assign work to be done out of sight?

We found that the amount of work completed is as important to the employee as it is to the company. "How am I doing in relation to others?" was a common if unstated question; it is one that is not asked when the "others" are just across the hall.

To deal with the issue Control Data has limited the kind of work done to those tasks with a definite beginning, middle, and end. Employees work on a project basis with defined milestones and a specific result or product to be delivered at a specific time. This means reliance must be placed on managers to be sensitive to employee needs for support, recognition, and structure. The manager must keep employees challenged at the same time he or she responds to their individual needs. The manager must be able to create an environment of caring, to communicate openly a concern for employees. Whatever abilities an effective manager has today will probably be inadequate tomorrow. The role of manager will no longer be that of sitting at the apex but rather that of strategic planning and consensus gathering, both of which put a premium on communicating effectively.

The successful managers in our alternate worksite program have good communications skills and a quality of caring. But more is required. In particular, it has been discovered that there is an even greater need for structure in this style of work and management than there has been in more traditional styles. People want to know what they are expected to do, and the manager must be able to communicate expectations of service and output effectively. One technique is to develop a contract between manager and employee for specific products or services within a set time; and the company is working on other approaches. It is recognized that we need to adopt a more formal project tracking system—an automated process for monitoring control—the implementation of which will be the responsibility of the manager. In the end we know we must be able to define productivity to the satisfaction of the employee, the manager, and corporate management.

After one year of the Alternate Worksite Program, Control Data has found that the productivity of the participating worker has improved. We have identified two major reasons for the improvement: First, with little or no time required in going to and from work, employees have more time to spend working; second, because working at or near one's home is desirable to many, their motivation is higher. In addition, because they do not have the stresses inherent in commuting, participants report feeling more refreshed when they start work. They also have fewer distractions and are able to work faster and with greater concentration. Accordingly, participants in one of the programs rated their productivity increases between 12 and 20 percent.

THE EMPLOYEE OF THE FUTURE

Human beings are social animals, and we do not always adapt well to the potentially lonely working environment of the electronic cottage. The fear of isolation is a real one, as is the fear of dehumanization by machines. There must be civilizing influences and a sense of community among workers, the sort of interplay provided by a joke in the hall or lunch with a friend. At Control Data this human need has produced an interesting phenomenon: underground, computer-based gossip. This electronic answer to the coffee break is an ongoing computer-stored series of conversations, jokes, games, non sequiturs, and running commentary. It can be called up at any time anywhere in the world, and it instantly puts an employee in touch with his or her peers.

It is important for employees to socialize and to have a sense of belonging. Therefore it is essential that those who work at home be included in meetings and regular functions with other employees. Control Data managers hold a minimum of mandatory once-a-month meetings with their "remote" employees. Many of these workers have an alternate worksite and a central office station, although most of their work is completed at the alternate site, be it at home or in a satellite office nearby.

As this trend continues offices will cease to be private and will become shared facilities with perhaps four desks and phones to serve eight employees. Not only will this lessen the demand for office space and facility expansion but it will also reduce overall corporate energy costs.

We realize that alternate worksites are not for everyone. Just as specific characteristics are required in a manager of these programs, so too the successful employee must possess certain attributes. He or she is highly motivated to succeed, tends to be more mature, and is a self-starter with little need for constant supervision. Several of the initial Control Data volunteers thought that they fit this profile but after several months found themselves becoming lax. They realized they needed the proximity of co-workers and the stimulus of an office environment in order to produce.

Employees can fall prey to the feeling that out-of-sight is out-of-mind, a fear that if they are not seen they will not be appreciated and that promotions might suffer as a result. Continuing encouragement and regular productivity reviews by management will help to mitigate these concerns.

The question of company loyalty has also been raised. So far this has not been a problem, perhaps because the people involved see themselves as pioneers and also see Control Data as a company that cares and is sensitive to their needs. Moreover, participants can claim some attractive benefits as a result. Some have cut their driving by 500 miles a month—a savings of as much as $90 in transportation costs. In California car insurance costs have frequently been lowered by 30 percent. Participants also save on food and clothing. An interesting exception to this provides an insight into the unexpected side effects implicit in any technological change. One of the participants used to take his lunch to work: During the noon hour he shut himself up in his office, unplugged the phone, and read. As an alternate worksite employee, he finds he needs to leave home at lunchtime to establish human contact and now spends more money eating out then he did before.

The very fact that alternate worksite employees are benefiting financially may become the program's biggest problem. The issue is that of "compensating benefits" for traditional employees who must still commute to work and arrange for meals away from home. Similar administrative problems are apparent in the areas of tax and insurance but there are still no concrete answers.

It is already clear that the success of such a program on a large scale depends on the manager-employee relationship. It is interesting to observe that the greatest obstacles come from managers rather than employees.

Those schooled in production tend to believe "You can't trust people you can't see." Unable to walk out to look over their troops, some managers experience "loss of stage" or fear of losing control. Because ego can play a large part in management, some managers equate a loss of visible employees with a loss of power.

The answer to these and similar problems is education. Managers can be made to understand that their responsibilities will be enhanced under such a system. In the short term they are seen as pioneers and given increased attention. Over the long term they will be able to supervise more people more efficiently. Use of electronic communications will enable them to reach more workers more frequently than is possible in person and with less chance of strategic misunderstandings. Further, if there is a problem in the system, it will come to light quickly.

Another source of managerial resistance stems from "keyboard fever," the fear of electronic technology. As in any fear of change, this, too, can be solved by education. As one learns more about the uses and potential of computers, fear of the unknown dissolves into a fascination with the possibilities.

Managers will come to realize the benefits of telecommunications, which can be used, for example, in an alternate worksite setting as a way of reaching group consensus, as a mediator, as a tool for performing tasks better suited to computer than employee, and for pulling together and analyzing data. Managers will learn to manage a process rather than just people, although the traditional managerial attributes of intuition and good judgment will always play a role. Managers will have to become more "computer literate," to understand and be able to access the ever-increasing volumes of knowledge stored electronically. Managers will have to learn how to supervise workers who are becoming increasingly specialized: As these specializations develop, the tasks of training, retraining, and integrating employees into the productive structure will become more difficult and more time consuming.

A NEW SYSTEM OF EDUCATION

It is in the areas of training and education that I believe human resource planning will grow most rapidly. Fueling the developments will be the cost of training, the explosion of specializations, and the resulting skill shortages. Some jobs will become outmoded; others will be created. The wide range of skills needed will require individualized instruction. To compete a company must, effectively and inexpensively, be able to train its new employees and replacements for its skilled workers.

The logical result is computer-based education and training that offers a vehicle for delivering both general and highly specialized courseware and instruction across a broad spectrum of need, from basic literacy skills to state-of-the-art technical advancements. Computer-based education is perhaps the only mechanism able to store, transmit, and deliver individualized instruction on a scale and with the speed necessary to meet the needs of business. Moreover, it can be made flexible and sensitive to accommodate individuals of widely varying backgrounds and abilities.

It was with these things in mind that Control Data developed the PLATO computer-based education system, which combines instruction on computer terminals with audio- and videotapes and textbooks. This system now offers courseware in fields as diverse as airline pilot training and small business management, but in each course the system is individualized to the extent that it proceeds at the student's pace, diagnoses precisely where the student's understanding has broken down, offers immediate reinforcement for correct answers, and assembles and stores a mass of information from which the student may take only what he or she wants. As the learning needs of employees become greater and more specialized, computer-based education has the capability of providing for them.

Another of computer-based education's assets is its adaptability to the remote-site environment. When Homework started, terminals were placed

Figure 3.   Stephan Masica works as a programmer at home; he also tutors homeworkers over the PLATO terminal.

in the homes of disabled employees who were then trained by computer. In this way a computer can meet the needs of the alternate worksite as an educational and training tool and as a means of production. One immediate result is the enlargement of the available labor pool, although the Homework program also accomplished some other things. From Control Data's point of view the program has reduced the high cost of maintaining disabled employees on medical leave and has given us a way of retaining productive and talented employees who would otherwise have been ruled out of our resource pool.

From the employee's standpoint Homework has led to feelings of intellectual stimulation and independence. Many people say the program has literally saved their lives and changed the outlook of their families. People with disabilities often keep unusual hours and may be found working or studying at their terminals at 3 A.M. PLATO has also been a counseling facility for them. Often isolated, participants who previously had few ways of expressing themselves have begun to share their feelings with one another over the computer network. A tremendous support system has been created and friendships have developed long-distance.

After the first year of Homework the twelve pioneering participants came to Minneapolis for a conference and banquet. The emotion of this reunion is difficult to describe, but people were deeply moved as one woman told the audience of Control Data management about her depression and despondency and then of the renewed self-worth and supportive friends that came to her through Homework. From a human resource perspective Homework has been a case of doing well by doing good.

CONCLUSION

In summary, the future of human resource planning lies with education and training. Management will have a greater challenge than ever in finding, training, and maintaining its work force. The background and makeup of this force will be extremely varied, and managers must respond with flexibility and caring.

Communication skills will be more highly valued than ever—as sought-after as MBAs were twenty years ago. Companies must be prepared to provide more for some employees and less for others. Lateral and even downward career moves should be anticipated as employees deal with stress and changing values.

Alternate work stations will gradually replace the central worksite, and employers must develop the skills and technology needed to manage employees dispersed to their electronic cottages.

We must become computer-literate. We must look for ways to understand, explain, and use the developing technologies. We must look for ways in which the corporation can help itself by helping its employees. Above all, we must remember that there is no limit to the possibilities of the human variable.

Note

Frank W. Schiff, *Looking Ahead: Identifying Key Economic Issues for Business and Society in the 1980s* Committee for Economic Development, December 1980.

12

THE PROFESSIONAL
ENVIRONMENT IN THE
TWENTY-FIRST CENTURY

LOUIS H. MERTES

As we plan for the future, we look for the things just emerging that appear to have perversity. I say "perversity" because emerging problems that are serious in nature will most certainly follow us to the twenty-first century. But then so will our new technologies. Improved, sophisticated, and adapted to new uses, they will become a resource that will help us combat the problems of tomorrow.

In attempting to determine what the professional environment in the twenty-first century will look like, this chapter focuses on the many and varied tiers of professional specialties that constitute the management of the large corporation with many domestic and international offices. Interest is centered on accountants, planners, economists, financiers, and salespersons, as well as on marketing and operations management teams. These people will be performing much the same functions in the twenty-first century that they do now—functions that by their very nature require a high degree of communication, negotiation, monitoring, research, and documentation. The need for information throughout this group is great as corporate strategies and objectives are formulated, executed, and evaluated. Disseminating the information requires crisscrossing the whole corporate hierarchy.

What then will be the biggest change affecting the corporate professional as we approach the twenty-first century? In my opinion it will not be what they do but how they do it. Let us take a closer look at some of the people in a typical corporation.

Let us take first a midwestern sales coordinator. This coordinator uses the study in his suburban home as his primary office. From here he or she directs all the activity in the ten small sales offices dispersed throughout the midwest in or near major industrial centers. He uses his telephone to dictate letters, memorandums, and reports to the corporate centralized word-processing center. The center relays his draft copies back to his electronic mail terminal at home to be edited and distributed throughout the day. By electronic mail the coordinator receives specifications for new contracts developed by the salespersons in the regional offices. He comments on, approves, or rejects them through the mail system. He works on his specifications at odd hours because he prefers to reserve the primary "office hours" for important customers and personal contacts in the community in which he is active.

The controller located in the head office is also a heavy user of electronic mail. He is responsible for the cash management function. Financial information from each office is instantly available to him and he disseminates it to his staff for further analysis. The controller also receives account balance and investment data over the mail terminal from his banks and investment advisory services. Up-to-the-minute bond quotes and rate changes are called into his telephone-answering device, which permits him to accumulate verbal data while attending to other matters. Key maturity or pay-down dates are stored in the mail data base with a follow-up date; thus the controller will be reminded by mail to act on these matters when necessary. Financial information, which is continually fed into several data bases and called up by the controller at will, is automatically displayed on graphs and comparison charts.

The operations manager is responsible for a production unit in a small town near the city in which the head office is located. Her unit had been moved about five years earlier to take advantage of a highly attractive labor market and lower operating costs. Although the operations manager uses the electronic mail and dictation features, she is more heavily dependent on her phone-answering mechanism than on any other technological tool. She carries a small device so that she can pick up her messages every thirty to sixty minutes while inspecting problem areas in the plant and consulting with her management staff. What used to be a small "beeper" is now a miniature recording and storage device; thus it is possible for her to relay and play back an entire message.

Those familiar with the recent explosion in microelectronic games and the popularity of large-screen TV know that none of what has been described is farfetched. All of these devices are available in some form today for office automation. Telephone-answering machines and equipment for dictation and word processing, for example, have been available and economically justifiable for the last ten years. What then will be the catalyst to force industry to embrace the technology that so far it has ignored or underused?

HOW TO OVERCOME RESISTANCE

Rapidly escalating costs at rates higher than the projected rise in revenue is the basic factor that is forcing businesses to address the issue of productivity. Because of the strong emphasis on automation at the clerical level in the last decade, significant strides or new opportunities probably will not occur again at that level. The professional level, however, has been virtually untouched. More than 50 percent of today's work force is made up of professionals. The professional level happens to be particularly affected by the high cost of energy because it is the most mobile corporate segment. Professionals usually commute farther to get to the office, travel to visit customers, and manage intraoffice activities or several offices. They are also heavily involved in community, industry, or government-related affairs outside the office. The costs generated by the professional, though necessary to support his continued contribution to the growth of the corporation, will become more visible—and more painful—as inflation and energy costs continue to soar. It is therefore safe to forecast that industry will ultimately push technology on the professional level in the same way that it eagerly adapted automation on the clerical level.

Inasmuch as the technology is completely or at least partly developed two questions arise: What are the obstacles that confront us and why is there not a greater acceptance of the technology today?

Two factors play a prominent part in the problem. The first involves the professional himself, particularly the professional who is well versed in his field and comfortable with his responsibility for other people. We professionals focus on devising practices and policies to make others more efficient; we are outward-directed and immersed in the issue of our staff's productivity. Deep emotional issues are stirred up when the issue of our own productivity is raised. All the clear and precise logic that we eagerly apply in automating the jobs and processes of the clerical work force becomes vague and confusing when turned inward on one's own habits and those areas with which we are comfortable.

Some executives may also be reluctant to give up what they perceive to be the accoutrements of their success: a secretary, a large travel budget, and several offices. Electronic wizardry does not yet convey the subtle messages associated with these older trappings of executive power. The executive realm of the corporation is steeped in tradition and status; its power symbols, though unconscious, are extremely difficult to alter or replace with new symbols.

The second cause of resistance arises because most professional people view their jobs as cognitive in nature. They describe their activities in terms of planning, forecasting, negotiating, researching, creating, and so forth. The motor tasks associated with the cognitive tasks—handling the telephone, walking to meetings, writing down data, sorting memos and inquiries—are largely subconscious and seldom analyzed. Yet it is here that tremendous savings and benefits in productivity can be achieved if we view many of our current tasks for what they are: wasted motor activity that reduces precious cognitive time.

Despite these hurdles, change in the form of automation and technological devices can be introduced successfully on the professional level now. In preparation for the twenty-first century, Continental Bank has for the last three years been installing large-scale office automation systems. A worldwide management inquiry system is currently used by 4000 managers and professionals. The electronic mail system disperses approximately 20,000 documents to 2000 professionals a week. These two systems alone make available to the managerial and professional staff more than thirty billion bytes of corporate data.

Several hundred telephone-answering devices, which handle up to 65 percent of the owner's incoming calls, have been installed throughout the corporation. Word-processing centers handle increasing amounts of telephone dictation. Remote word-processing centers are being tested and evaluated in two modes: in neighborhood suburban centers and employees' homes. In each case the employer is able to tap an attractive labor pool that was previously inaccessible.

MAKING USE OF EARLY ADAPTERS

The personality profile of the professional is not the only cause of industry's slowness in introducing information technology successfully on the professional level. Another cause is the failure on the part of the corporation to market and package these innovations creatively—in other words, to sell them effectively.

In today's world the people easiest to reach and impress are those members of the population who are eager to be first to try anything. It also happens that the professionals of the twenty-first century are today's children, who are playing with electronic games and simple hand-held computers. These children also use minicalculators in school. The world of electronics will be a familiar environment for the young men and women of the twenty-first century; they will have an adeptness and an acceptance level far beyond that of most mature people today.

Some people are by nature early adapters—quick to see the benefits of new products and eager to incorporate them into their environmental structure. They are tolerant of imperfections because they clearly see trade-offs in accepting the negative features for the sake of positive ones. Implementation programs designed to involve these early adapters stand a much higher chance of being warmly received and enthusiastically supported. An early wave of product acceptance is often sufficient in and of itself for winning over the next group of staff members—those people who will warm to a new product as soon as someone else pronounces it valuable or worthwhile.

Continental Bank has used this approach to introduce automation at the professional level. Small pilot programs with minimum funding and inexpensive equipment, staffed by a few highly enthusiastic people, were set up wherever the environment seemed receptive. Several pilot projects were started simultaneously, each using different approaches and different population segments. The success or failure of each was evaluated carefully. Because the allocation of capital resources was small, it was unnecessary to force success or to postpone the natural death of failure. Strategies could be altered easily, and the process remained dynamic and flexible.

Pilot programs need not encompass an entire working group; they can be targeted at various individuals throughout the corporation. Some systems and technologies, such as a data storage and retrieval system or computer graphics, can be used effectively by an individual for his or her own benefit. Information retrieval has a particular virtue as a pilot project: It requires perhaps the least behavioral change by all parties. If the user is able to obtain information on paper or terminal soft copy on demand, the system can be treated as if it were a paper-based filing system. An assistant can obtain the information personally. Individual early adapters can be used for this pilot with little concern for the prospects of eventual total implementation throughout a work unit. It may also be possible to use existing terminal equipment, thus minimizing the cost of the pilot.

If there is no extra equipment cost, the pilot will be under less pressure to demonstrate benefits too early in the learning curve. Later, when the benefits become clear to the early participants, others will become interested. Integration into the work environment will proceed on an individual basis with no need to mandate use until well into the project. It may be necessary eventually to mandate use for specific purposes to eliminate production and distribution of older forms of information.

Graphics information is probably the next easiest technology to implement, and once again early adapters can spearhead the pilot project. Because of the unlikelihood that existing equipment can be used, the cost of the pilot will be increased; this adds pressure for early demonstration of benefits. It is important, therefore, to identify the clear advantages of the pilot project to the people involved before it begins. Anything less than an enthusiastic reception will endanger its success.

The next stage in the development of an electronic information system is word processing itself, which requires both organizational and behavioral changes affecting a whole group rather than individuals. Productivity increases when several people are served by a single word-processing operator. This shift from "dedicated" to "shared" typing causes a fundamental change in the work environment and creates many conflicts. Fortunately, this transition has been made by many firms, and the newcomer has available valuable information from many sources.

The introduction of word processing quickly leads to better management of the logistics involved in supplying work from several people to the word-processing operator. Telephone dictation is one effective way to do this, but it requires that new skills be learned by an entire working unit—not just by the early adapters. There must be thorough training at the start of the pilot project and close follow-up of each person. Once the group has adapted to the technique truly remote telephone dictation becomes possible, and at this stage word-processing operators or their principals can actually start working at home.

AUDIO AND ELECTRONIC MAIL

Audio mail requires the same skills that are used in telephone dictation. In both cases the user must present a complete thought with no response from the other party. Individuals, of course, can install phone recorders for their own use, but the real benefit does not develop until all the people in a working unit have the device and use it properly. If the normal verbal communication pattern of a peer group is to be covered effectively, a pilot project must have approximately twenty units. (If a pilot project is too small, there will not be enough nonsimultaneous transactions to justify the project.) There must be adequate training at the start of the pilot to get people to leave complete messages and return calls with complete messages. Because the entire interacting group must participate to achieve success, there must be careful follow-up to ensure proper use of the system. This monitoring must be done by the early adapters within the pilot group itself; no one else is available to do it at this stage of development. Initially many people will have negative comments about talking to a machine. It is up to the participants, not the trainers, to explain the time savings possible to the callers; otherwise many people will simply hang up or just ask to be called back. They must be encouraged to leave a full message so that the transaction can be closed with the return call.

Perhaps the most difficult technology to introduce and institutionalize is electronic mail. Some users will regard it simply as a delivery vehicle and have someone else pull a printout of incoming mail and key in outgoing mail as if it were typed and conventionally mailed. Eventually most people will discover that it is faster and more productive to read their mail on the terminal, and almost all users will soon begin to key in one- and two-line comments. In fact, a new problem will develop for a few people who key *all* their own correspondence. These people must be encouraged to use telephone dictation for their longer documents in the interest of greater productivity.

As with dictation and audio mail, electronic mail will require that new skills be taught to an entire work unit, but even more is required because written material will characteristically move through several layers of the organization. By cutting vertically through the organization the pilot project is exposed to resistance from at least one and perhaps several high-level people who classify as slow adapters and therefore present a situation that requires careful planning and execution. The pilot should have early support from high-level, early adapters but should allow sufficient learning time for the late ones. Individual training and follow-up are also needed to avoid unnecessary interlevel organizational pressure.

Even when fully developed, a system of nonsimultaneous communication will not eliminate the need to meet face-to-face on some issues. In these instances teleconferencing can save time lost by traveling to many of these meetings. Picturephone services will be available not only between cities but between a firm's various offices within a single city.

Experience has shown that creative experimentation followed by careful evaluation of the results is the appropriate way for a corporation to introduce methods that help professionals become more productive. Armed with a working knowledge of the available tools, the executive can choose those best adapted to the purpose and can determine the most effective means of introducing them. The executive can also circumvent psychological barriers by encouraging each professional to develop his or her own unique style and method of participation. The professional who discovers that productivity has increased will encourage others to try the same methods.

LOOKING FOR MODELS

The overall message is that electronics increases productivity by releasing individuals from the need to do their work simultaneously. People do not have to be in the same place at the same time to work efficiently; they can proceed at their own pace, regardless of the schedules of others. The time previously wasted in the logistics of synchronizing work can be devoted to far more creative functions.

In light of Continental Bank's experiences, I believe that the manager-professional of the twenty-first century will depend heavily on microelectronics to remove artificial restrictions on mobility and communication. Professional energies and skills will be focused more sharply on the real cognitive tasks at hand; the creative manager, by using the appropriate tools, will outperform the traditional manager.

Considerable adjustments in personal style will have to be made, of course. As all information workers move toward the nonsimultaneous mode, the familiar management techniques of observation will no longer be adequate to control workers and events separated by time and space. Large central offices will change as fewer people use the facilities for face-to-face meetings and as the workday lengthens with fewer people present at any one time.

As the centralized office loses importance, we must keep in mind the social as well as the business character of the office. Many people will be unwilling to forego the social aspects regardless of the various advantages inherent in working at home or in being highly mobile. In some cases the seeming advantages of working at home may turn out to be illusory; for instance, it could be difficult for a single person living in a singles' apartment complex. Similar problems may face the married person who would be trading social contact with professional peers during the daytime for the companionship of adults who do not work.

Because people have different social needs in their private and professional lives, we can expect that in general the evolution of the professional workplace will follow much the same evolution that has taken place in the medical world.

Just as hospitals provide a high concentration of professional interchange, advanced techniques, research, and teaching, central offices will continue to perform comparable functions in all industries. At the neighborhood level professional buildings and motels will be used in much the same way that the medical profession uses clinics. In this environment small numbers of professionals can maintain their peer and personal contacts. Finally, there will be individuals who choose to work out of their own homes just as doctors engage in private practice.

In this new professional environment on-the-job training will be much less effective than it currently is in the office. A professional apprenticeship program may be required for those who choose to be highly mobile. A new employee might go through an orientation and training procedure that involves traveling with an experienced individual. During this time he or she will learn how to use mobile communications effectively and how to structure time. For those who are inclined to work at a fixed location there may be a residency requirement at the central office much like the residency requirement of the medical profession.

Regardless of whether managers and professionals in the twenty-first century choose to stay put or be highly mobile, they will have to learn new skills. Perhaps it would be more accurate to say that some older, neglected skills will have to be revived. At the present time training in interpersonal skills is oriented toward meetings and negotiation. There will be fewer occasions on which this kind of interaction between people will take place. The effective use of dictation equipment, phone recorders, pagers, and electronic mail requires the presentation of a clear, concise message that leaves no opportunity for the recipient to ask questions immediately. The skills needed will be those that existed before the era of the telephone when most business communication was done by letter. They must be revived in the new era when most business will be conducted by verbal mail.

When workers do not choose to use the central office, the firm must provide for the social aspects that are lost. There is a need for formal identification with the firm and the unit, a problem that sales organizations have solved with the periodic sales meeting. Individuals will be required to come together on a weekly or monthly basis. Identification with the firm and its goals can be reinforced. At these meetings people will also experience identification and interaction with their peers.

A concept such as the "quality circle," now used widely in Japan, may provide an answer to this general problem. Companies in the United States are now beginning to experiment with quality circles and with good results. A small group of people who share a common work experience meet regularly to discuss issues and problems that affect quality. They are trained in various group dynamics and problem-solving techniques to facilitate the process of offering recommendations to management. These circles build rapport and involvement among co-workers and at the same time provide access to top management in a structured approach. Concepts like these will become increasingly important in a dispersed office to maintain loyalty and a sense of purpose and direction toward the corporation.

Corporations will also make an effort to encourage their people to join neighborhood organizations, such as homeowners' associations, school boards, and local service groups. This activity will not only replace the old workplace social groups but will also establish the firm's presence in a community. As this transition takes place from office social contacts to neighborhood social contacts, firms will recognize the need to give financial support to their workers' membership in professional and civic groups and local country clubs.

The business environment that we have described for the twenty-first century represents an evolutionary development from that of today. No revolutionary breakthrough is needed to bring about the future office. A slow, steady, creative process of change will take us into the twenty-first-century environment. Many firms will make incremental changes in office equipment and procedure—and discover after the fact that they have arrived there. But they will have arrived later and with less benefit than those who set their sights on the future and then control and direct the evolutionary process. The task at hand is to get started with the tools that are now easily available.

In summary, present technology can help us to proceed toward the professional workplace of the twenty-first century. Other professions, such as medicine and sales, have gone down this path before us; they have provided us with models of organization. We need only to take advantage of the infrastructure we already have in place. In some cases new equipment will be necessary, but it will be easily obtained. Beginning the creative process of testing and experimenting with tools to liberate the professional from tedious office tasks is the only major hurdle. If we believe that automation by the use of microelectronics has its place in the professional environment, and I firmly believe it has, the time we waste today by not beginning the evolutionary process can only detract from our competitive edge in the workplace of tomorrow.

13

TELECOMMUNICATIONS AND BUSINESS POLICY

PETER G. W. KEEN

No field that encompasses "telecommunications and business policy" has yet been established. Discussions of the impact of communications technology usually focus on hardware, public policy, and regulation or on such specific applications as office automation, teleconferencing, and electronic banking. A more integrated approach is now needed. The emergence of a multitude of technological building blocks—computers, information storage devices, terminals, and communications links—makes it possible to develop new markets, add new services to those existing, and redesign organizations and work patterns. It is critically important to view them in a larger context to evaluate the opportunities and problems they present.

This chapter accepts the truism that communications technology presages a new industrial revolution that eventually will reshape our society. To forecast when and how this will happen, however, would be foolhardy. In general, technical change far outpaces social change.[1] Institutions adapt

The author acknowledges the contribution of Matthew Devlin and Richard Mills of Citibank N.A. to the ideas expressed in this chapter.

slowly. The interaction of technological advances with unpredictable political, social, and economic courses may dampen rather than accelerate the effects of technology. The last thing we need, therefore, is another gee-whiz fantasy to describe the information society, the office of the future, the communications center in the home, or the cashless society. Yet fanciful as these may be they do indicate the major *direction* of the change that is occurring.

The assumptions on which this chapter is based are conservative. The view taken is that organizations will try to assimilate the new technologies into their existing procedures and structures as they go. They will adapt to the new technologies cautiously by avoiding high-risk ventures but exploiting opportunities sensibly. To a large degree the existing investment in and uses of computers will constrain future innovation.

Four factors in particular will be critical when organizations plan a business strategy for telecommunications.

First comes the development of new products and services made possible by compunications, a term mentioned earlier in this book that refers to the combination of computers and communications.[2]

The second factor is the delivery base that involves the integration of on-line processing and enquiry, an accessible data store, and terminals as close to the customer as possible. Telecommunications is the infrastructure of this capability.

Third is the design of organizational mechanisms and structures. In thinking about this step, we must bear in mind that we are moving toward a world in which spatial and physical constraints will no longer determine organizational arrangements.

Finally, there is the impact that telecommunications will have on the nature of work at all levels—clerical, administrative, supervisory, and managerial—as it becomes more computer-mediated, to borrow the concept developed by Harvard Business School professor Shoshanna Zuboff.[3]

Even under conservative assumptions, communications technology will radically change the structures and processes of most organizations well before the end of this century. The manager's job will require many new skills in terms of planning, control, and supervision, and we can expect some broad changes.

Competitive advantage in service industries will depend heavily on compunications; this is already occurring. Organizations will be designed around their telecommunications systems, which will permit decentralization *and* centralization at the same time and will significantly alter planning, control, communication, and reporting mechanisms. The first stage in this process can be seen as the roles and influence of corporate staff are changed by compunications. As more and more of the organization's activities are carried out by compunications, work will become more abstract.[4] This will also be true in the organization, especially if teleconferencing becomes a substitute for travel or clerical work centers become geographically dispersed.

Correctly or incorrectly, organizations are assuming that necessary productivity gains can come only from long-term automation of office and administrative functions; jobs will be specified and groups formed explicitly in reaction to the evolution of compunications.

TELECOMMUNICATIONS AND INDUSTRIAL MARKETS

The United States has become a service economy.[5] Compared with manufacturing, which is generally characterized by economies of scale, service industries tend to suffer from diseconomies. As volume increases a disproportionate rise in activities is needed for coordination and administration. Organizations then require some means of improving effectiveness (revenues) or efficiency (costs). Computer-based innovations obviously offer the potential to do both when properly managed. Business is beginning to look on computing technology as a strategic rather than a tactical investment, one aimed at efficient processing of transactions.

Computers can improve revenues by upgrading the quality of existing products by a reduction in errors and turnaround time, by providing new products and services, and by leveraging the skills of key individuals with planning and decision-support systems and with market and environmental information.[6]

Computing can improve costs by cost displacement, cost avoidance (e.g., handling an increased workload with no increase in personnel), and office productivity. In speaking of productivity in the office, the phrase often used is, "paper is the enemy."

Investment must also be made in "base-building," which provides the infrastructure that is needed to achieve the goals outlined above. Base-building requires the development of a data management and telecommunications capability, pilot projects to test new technologies, and educational programs, especially those needed to reduce fear of computers and the culture gap between users and technicians. Table 1 offers illustrations of these types of investment.

Computers at last have become an opportunity instead of a problem. Disaggregated data, stored at the detailed level needed for processing and inquiry and combined with an economic, powerful telecommunications network, are becoming the key to competitive advantage in service industries.

Consider banking. It is difficult to think of any new, highly competitive banking product or service that does not depend on telecommunications: Pay-by-phone services, cash management systems, electronic funds transfer, and automated tellers come immediately to mind. The recent incursions of Merrill Lynch into what has traditionally been the banks' marketplace are based on its network. Reuters, American Express, Dun & Bradstreet, and others are pushing new services closer and closer to the customer.

Virtually all major organizations, financial and otherwise, are moving toward the integration of the technical building blocks shown in Figure 1. The capability illustrated here is an ideal; it will be a long time before all the necessary components are available to achieve it. Software for data-base management, for instance, is not yet adequate for high-volume, high-activity file processing. What the schematic represents is the logical, comprehensive capability needed for utopian visions of the information society and office of the future to become real.

Communications and the data store are the keys to customer service. The addition of an immediate-transaction processor provides the customer with the fastest possible turnaround. When a transaction is made, the company's books are updated at once. This combination of facilities—for example, in on-line systems for order entry or electronic banking—means that the gap between the customer and the organization is dramatically shortened in terms of time and space. The terminal provides direct access to services. In effect, the organization reaches farther and farther into its environment and takes the customer inside.

This process is far-reaching in its consequences for operations and innovation in intra- as well as interorganizational activities. Decentralized activities can be coordinated as if they were centralized.

Table 1. Examples of investments in computing

Goal	System	Technology
Improving revenues		
Existing services	Customer service stations	Disaggregated data store (transaction level) and terminal access
New services	Network information services	Communications network and data store
	Corporate cash management services	
	Nationwide real estate information services	
Leveraging productivity of managers	Decision support systems	Stand-alone or time-shared computers
	Planning languages	End-user languages
		Graphics
		Information access tools
Improving costs		
Cost displacement	Automation of manual and clerical procedures	Mainframes and data capture tools
Cost avoidance	Personnel inquiry and reporting systems	Data store and simple end-user software
Office productivity	Text handling distribution and storage systems (far broader than word processing)	Integrated, standardized network
Base-building		
Telecommunications	Highways for application vehicles	
Data management	Central, accessible resource for services	
Education	User is active component in on-line system. Management has proactive role when compunications are key to market strategy and productivity.	

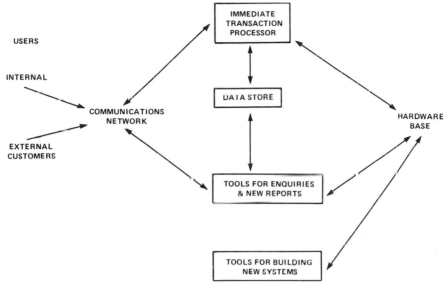

Figure 1. The ideal capability.

Immediate Transaction Processor.   Equivalent to a perfect clerk; software for on-line order entry, cash management, etc.

Data Store.   Logically, a centralized, accessible library of data regardless of physical structure and location.

Tools for Inquiries and New Reports.   Packages and nonprocedural "end-user" languages that eliminate dependence on professional programming staff.

Tools for Building New Systems.   Systems development languages and productivity aids.

Hardware Base.   Configuration of mainframes and/or minis; may be centralized, decentralized, or distributed (using communications links).

Communications Network.   Voice and data network linking users to the transaction processor data store and inquiry software via the hardware base.

THE PACE OF CHANGE
Incubation and Takeoff

The "ideal capability" just pictured describes a likely convergence and integration of technologies that already appear to be taking place. Progress toward this state is already substantial and is accelerating, as shown by innovative developments over the last few years in banking, which in most cases involve standard transaction processing and data storage.

Perhaps the most dramatic development has occurred in international banking (Table 2). Electronic check-transfer capabilities have improved to a point at which same-day settlement of accounts between international banks was scheduled to begin in the fall of 1981. Once again, no innovations in procedures or technology are involved, yet the consequences are immediate. Formerly, settlement was made at 10 A.M. the next day. The new Clearing House International Payment System (CHIPS) now puts an end to the existing sixteen-hour float, which is more than $1 billion. Banks are having to change many of their operations: Weekend arbitrage, overnight Eurodollar deposit rates, and loan agreements are all affected. Here is one comment: "European bankers were stunned when the full implications of the change were explained to them."[7]

A key aspect of this development is that it uses a combination of technologies that have been available for a long time; this is also true of similar developments in other industries. Videodiscs, satellite communications fac-

Table 2.   Developments in international financial market services

Information services	Reuters Monitor and AP/Dow Jones Telerate provide up-to-date rates in foreign exchange markets and international news services
Computer-supported dealing services	Permit "free dialog" trading; dealers converse with ease with others via CRTs, Reuters Money Dealing service, and EUREX
Computer-matched dealing services	Match bids and offers; ARIEL executive service and EUREX system for international bond market
Automated settlement services	Euroclear and Cedel for Eurobonds; London Stock Exchange Talisman service
Funds transfer and message systems	SWIFT links more than 700 banks in North America and Europe

simile, voice mail, OCR (optical character recognition), among other devices and systems, will only speed up these developments and extend their variety and innovativeness. Communications technology is the infrastructure for such applications as office automation, new customer services, electronic banking, network information systems, and teleconferencing.

Considerable time is required to put an infrastructure in place, but once it is there its exploitation is rapid and almost inevitable. This is an important point. There is a paradoxical aspect to the pace of change in information systems. The frequent lags between technical and social or organizational change in early stages of growth are associated with rapid innovation at late stages. When a new technology appears, we expect it to change faster than it does. An infrastructure needs to be built first. Later we assume that the slow pace of change will continue and, once the infrastructure is in place, we are surprised at the sudden acceleration.

This paradox can be explained in terms of Richard Nolan's concept of the stages of growth of an organization's data processing, which is probably the most influential conceptual framework in the management information systems field.[8] He argues that data processing moves through an S-shaped growth sequence. He initially defined four stages but added two more to reflect a shift from computer management to data resource management.

Nolan presents a relatively monolithic view of technology. He does not exclude communications explicitly from this framework. He seems to view communications, as do many data-processing managers, as a tactical ancillary to the other components of the ideal capability rather than as the strategic integrator that makes those components a new competitive force. The main insight behind Nolan's concept is that complex technologies involve a learning curve. At any given point different technologies may have reached different stages in the learning curve, and the slope of the curve may vary; for example, personal computing had a short germination before it really took off; it is too early to tell how steep and long the growth will be before it reaches a plateau. Telecommunications, data-base management, and office automation involve long lead times, significant organizational changes, and resistance (or at least cautious concern).

The S-curves for a hypothetical organization may look like Figure 2. Any technology can be mapped in this way. The key questions are these: How long is the incubation period and how steep is the take-off thereafter? The more complex the technical and organizational infrastructure needed, the more likely it is that gee-whiz prediction will be too optimistic—the

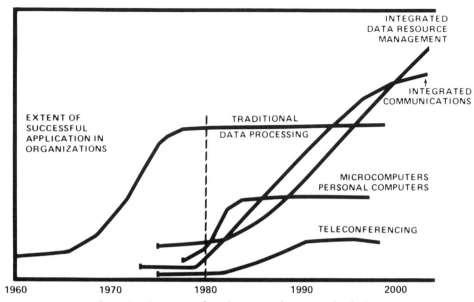

Figure 2. Concurrent learning curves for new technologies.

cashless society in 1975 was seen as just around the corner. When the infrastructure is in place, the predictions may err in the other direction.

Telecommunications is the infrastructure for many new applications: office automation, new customer services, electronic banking, network information services and teleconferencing, for example. There is plenty of evidence that we are now beginning the steep slope of the S-curve. We may be taken aback by just how fast and immediate future change will be.

WHO IS THE "USER"?

As long as an organization's use of computer-based technology involves separate components, it can be managed by traditional mechanisms. These separate systems include transaction processing, data management, and the configuration of the hardware base for internal purposes. The data-processing function is responsible for meshing the components with the company's activities.

All this changes when Merrill Lynch presents a radical challenge to the banks with its Cash Management Account, Chemical Bank introduces a corporate cash manager service, Citibank makes an all-out drive to capture the consumer market in credit cards, or American Express acquires a brokerage house. The business and technology plans must interact. Top management has to be more proactively involved in issues that relate to technology instead of making go or no-go decisions about single projects. The focus on marketing and customer services requires that management look beyond the technology to the environment instead of concentrating on the relatively parochial concerns inherent in intraorganizational processing.

In the earlier discussion of the ideal capability one key component was omitted: the users. In batch processing, users have a passive role; if they fill out forms correctly, the computer system will generate the correct results. In an on-line environment, however, the users become an *active* component. Their skills, attitudes, and initiatives determine the quality of the system.

Until now the primary users of computer systems have been clerical or data-entry staff. Managers, other employees, and customers, as secondary users, receive reports, invoices, and checks. Communications pushes the computer into new cultures. The customer directly interacts with a bank's processing systems through corporate cash management services. Electronic message and inquiry systems, text-handling tools, and network information services involve hands-on, active interaction by nontechnical individuals who have been buffered from the computer.

Figure 3 shows the extent to which the communications infrastructure adds new classes of user to an existing computing capability. Primary users interact directly with the computer resource. Those outside the circle must locate a primary user or an intermediary; for example, without a public network customers must interact with the customer service staff who, in turn, work with the accounting group, programmers, or data-entry function if there is no internal network. Figure 3 implies a substantial cultural change and major shifts in organizational and managerial mechanisms.

The tidy, closed world of traditional data processing is being opened up. It seems fair to assume that there will be no single functional area or job level that will *not* be pulled into a direct relationship with computing. A substantial culture gap exists between computer people and the rest of the organization, and even more culture shock will be felt as communications changes the nature of work and the organization.

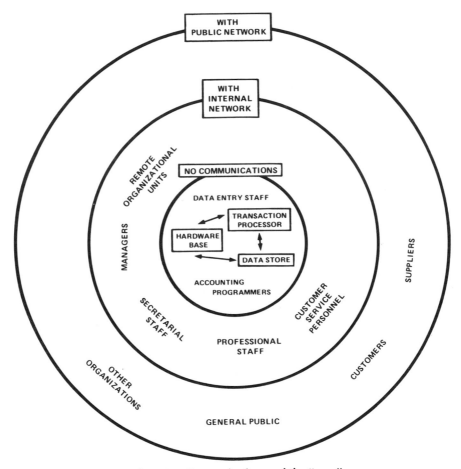

Figure 3. Communications and the "user."

TELECOMMUNICATIONS AND ORGANIZATIONAL DESIGN

In the late 1960s a multinational company, concerned about the need to coordinate its European affiliates, created a planning group in the capital city of its most profitable subsidiary. The affiliates still operate in a decentralized mode. The planning unit is widely regarded as an evil, intrusive and unresponsive even if necessary. The local subsidiary finds it extraordinary that its contacts with the United States are fewer and slower than before. Many of its personnel resent the superstructure of American managers that now gets between them and the marketplace.

It is ironic that the only reason for creating the new unit in the first place was to speed up and coordinate planning and communication. It was a clumsy solution. At a time when no internal telecommunications network was available the desired goal could be achieved only by putting a new organizational unit in the right location.

In general, the design of organizational structures has been based on physical arrangements. Centralization and decentralization are seen as necessarily in conflict. Time and space constrain design. Centralization sacrifices time to gain control; decentralization involves the reverse. Many corporate staff and reporting functions are purely devices that provide message switching or information pooling.

The organizational design implicit in Figure 3, when all three circles are in place, makes many of these devices superfluous. The multinational can now use its communications network to coordinate the activities of decentralized units. It may even choose to *increase* the autonomy of the affiliates because it can monitor their activities and respond to changing situations far faster than before. Daily reporting of a few key figures can replace the more detailed monthly feedback. The organization can have responsiveness *and* control. Is this an increase in centralization or decentralization? The dichotomy becomes meaningless.

There is no systematic study of the relationship between telecommunications and organizational design, but there is empirical and theoretical support for the view that organizations will be structured around their communications networks. The role of corporate staff shifts rapidly whenever there are changes in control of data and provision for shared access to planning information. The link between information and influence has often been noted; data are a political resource and building a data base is often a political act.[9] In Figure 3, the "owners" of data are the primary users; others must often go through them to talk to one another. Much of their influence is eroded with changes in the communications infrastructure.

The outer ring in Figure 3 represents an entirely different combination of procedures and authority than the inner ring. Common sense indicates that if the outer ring were created to provide new services to customers and to build a competitive advantage resources and job functions would be shifted to support them. We can expect that the activities critical to effective development and use of the ideal capability in Figure 1 will report at a more senior level than before and be backed up by more professional staff—in terms of numbers and sophistication.

This is happening already in many organizations. Information is revealed as a corporate resource and the information function is thus analogous to the corporate financial staff. The data-base administrator then needs authority, not just technical expertise. The head of information systems reports directly to a senior operating manager and frequently to the CEO.

Most people view organizations as pyramids on which position is marked by hierarchical status and horizontal function. Telecommunications not only allows us to see them more abstractly as systems for information processing but it becomes the vehicle for major redesign. Already it permits centralization-with-decentralization, shifts the influence of central staff, eliminates spatial constraints on organizations, redistributes data, creates new couplings between previously separate units, and blurs the boundaries between units and between the organization and its environment. That adds up to a radical set of forces for change.

THE IMPACT OF COMMUNICATIONS ON WORK

There is little question that communications technology increases the abstraction of the organization and the job. Going to work and handling paperwork are concrete, whereas almost every trend in compunications contributes to more abstraction.

What was once concrete—a form, money, a letter—is now intangible, even invisible. The CRT comes between the worker and the task. This is not necessarily threatening or harmful to people, but it is different. In an ongoing study Zuboff raises many questions about the psychology of computer-mediated work, but most of them remain unanswerable.

The one certainty is that paper is management's avowed enemy, and in the name of productivity managers are committed to defeating it. Economic forces thus ensure that more and more of the work done by paper will be handled by word processors, electronic message systems, facsimile, and microfilm. Every manager knows that technology costs are dropping at a compounded rate and white-collar labor costs are rising.

A typical estimate of the decrease in technology costs per year is 20 percent and of the increase in labor, 10 percent. That means that even if computers are now twice as costly as manual procedures for a given task the crossover point is only two years away. Even if technology is three times as costly and decreases at only 10 percent a year, the break-even point is four years.

By no means does everybody see it this way. J. L. King and K. L. Kraemer stress the hidden costs of decentralized computing, particularly the organizational resources needed to manage an increasingly complex system that is vulnerable overall to malfunctions of a single subsystem. They point to previous failures to anticipate the interdependencies and dependencies that new technologies introduce, for example, in the use of energy, automobiles, and chemicals. "The current claims for decentralization potential and productivity payoffs from implementation of telecommunications technologies," they say, "rest on only the loosest understanding of the social, technical, and organizational dynamics that surround actual technological adoption of routinization."[10] Others agree, including Zuboff, who believes that the "automation of managerial assumptions" may be costly indeed.

Office automation is nevertheless taking place, and the pace is speeded by economic stress and the tremendous emphasis on productivity. Implicitly, such a process will be costly to many workers. It may be so for managers and supervisors, much of whose authority and control is based on physical location. New workers are socialized partly by the physical surroundings and the organizational climate they create. If the organization becomes more abstract, so too do many of the traditional tools of management. The substitution of communications for commuting and travel may or may not involve costs and stress for workers. It certainly will make new demands on supervisors. When people work at home or in neighborhood work centers, where is the "organization"? How does one motivate, control, and evaluate the distant employee? How does one manage meetings by teleconferencing? Many of these questions are considered at greater length by other authors in this book.

Communications technology, even in its present form, is forcing consideration of major choices in market strategy, organizational design, and the education and management of the work force. Failure to plan will be expensive; an inadequate network constrains innovations in services and reduces the range of options for organizational design. As the communications infrastructure becomes more and more central to both external and internal operations, the business plan not only will have to drive the technology plan but also will be driven by it.

There are relatively few experts in the strategic technical, regulatory, public policy, economic, and organizational aspects of telecommunications. Expertise in the business policy issues is almost nonexistent because an organizational, business, and technical perspective is involved. At the very least an organization must recognize that its communications network is far more than a set of voice and data lines. It is an abstraction. In this it resembles that longer established, better understood strategic resource of the organization, money, which is both "real" and "funny," centrally allocated and monitored, and decentrally used.

Too often at present the telecommunications function consists of technical specialists in relatively junior staff roles. Communications are seen as an extension of computers. These views are myopic.

The topic of telecommunications and business policy is as important as that of corporate financial management. The chief financial officer is not a staff bookkeeper isolated from the capital investment cycle. The chief communications planner has to sit at the same table. At a time when the incubation period for telecommunications seems to be almost complete, that planner needs to get ready for a period of complex, challenging change that will affect almost every aspect of the organization. Although we cannot predict that change, we can prepare for it.

Notes

1. Peter G. W. Keen, "Information Systems and Organizational Change," *Communications of the ACM*, **24**(1), 24–33 (January 1981).
2. Walter A. Hahn, "The Post-Industrial Boom in Communications," in C. Stewart Sheppard and Donald C. Carroll, Eds., *Working in the Twenty-First Century*, Wiley, New York, 1980.
3. S. Zuboff, "Psychological and Organizational Implications of Computer-Mediated Work," Center for Information Systems Research, Sloan School of Management, MIT, Working Paper No. 71, June 1981.
4. Zuboff, *op. cit.*
5. Eli Ginzberg and George J. Vojta, "The Service Sector of the U.S. Economy," *Scientific American*, **244**(3), 48–55 (March 1981).

6. Peter G. W. Keen and M. S. Scott Morton, *Decision Support Systems,* Addison-Wesley, Reading, MA, 1978.

7. "What's Happening in Banking in Europe: Predictions for the Next Decade," *Datamation* (March 1980).

8. R. L. Nolan, "Managing the Crises in Data Processing," *Harvard Business Review,* 115–126 (March–April 1979).

9. Keen, *op. cit.*

10. J. L. King and K. L. Kraemer, "Costs as a Social Impact of Information Technology," in M. L. Moss, Ed., *Telecommunications and Productivity,* Addison-Wesley, Reading, MA, 1981.

V

INTERNATIONAL
PERSPECTIVE

The final section of this book deals with what many consider the knottiest problems of the new age of communications, at least in political and confrontational terms. These problems have to do with the vast flows of electronically generated information that now move not only within countries but across national boundaries as well, from one country to another and from one continent to another. Individuals, businesses, and governments feel strongly that they have a vested interest in maintaining access to this information and in having a say in how it is originated and distributed. Various highly complex issues, which range from the potential invasion of personal privacy to seeming threats to national sovereignty, are involved. By presenting very different viewpoints the four contributors to this section of the book provide important insights into some of these complex problems.

Harry L. Freeman asks how multinational corporations will fit into the long-term communications picture. In framing an answer, he makes an often overlooked point: The fastest growing multinational corporations of the world today are not European or American but the state-owned petroleum companies of OPEC. These MNCs and the Eastern Bloc state-owned trading companies will be part of the international communications industry in the future—an industry that will be characterized by giant, competing conglomerates of various types and under varied ownership.

Kimio Uno asks how well society is prepared to deal with the many issues raised by the extraordinary worldwide development of information and communications networks. He reports that in Japan interdisciplinary institutes in the universities are deeply involved in experimental programs that are dealing with communications and information processing. In one such project an attempt is being made to develop interactive social behavior systems to explore the group dynamics of the democratic process. Another project has been designed to build a socioeconomic simulator and data base that will be used by business, government, unions, and other groups for thrashing out major national policy issues.

Herbert I. Schiller wonders how long the free flow of information can survive the pressures of three-quarters of the world's population, which is seeking an increased share in global and local resource allocation. His position is diametrically opposed to that taken by Congressman Wirth: "As in the past, cultural as well as economic, technological, and scientific developments will be shaped by popular movements." In effect, Schiller is restating the power issue that underlies all discussions of electronic communications technology.

Finally, Kaarle Nordenstreng questions some widely accepted basic assumptions. How valid, he asks, are the concepts suggested by the terms "information society" and "global village"? These visions, he suggests, are ill-founded and misleading. "National frontiers are not necessarily withering, nor are societies inevitably being pushed by technology into a relatively homogeneous 'United States of the World.' "

14

THE LONG-TERM OUTLOOK
IN COMMUNICATIONS

A Corporate Executive's View

HARRY L. FREEMAN

This chapter attempts to set down in some orderly form the concerns of senior corporate communications executives as they try to peer into the future of the communications industry. Needless to say, this is a vast subject filled with uncertainties. To get a handle on it I find it useful to ask four questions that have special pertinence. Each question seeks to isolate a key issue that will determine the outcome of the competition in communications that is developing on a global scale as we head into the twenty-first century.

First, where are the government-owned or regulated postal, telegraph, and telephone companies of the world going? These organizations are familiarly known in the industry as PTTs.

Second, where, as institutions, are the multinational corporations going and how will they affect the communications picture?

Third, where is the technology of communications going?

Fourth, as a result of this technology, what companies or institutions will be our long-range future competitors?

To be aware of the particular insights or biases I have brought to this inquiry, it may be useful for the reader to know my position in the industry.

As a corporate officer, my responsibilities lie in two general but overlapping areas. In the first I chair a corporate-level strategic planning group. I say "corporate level" because American Express has many operating units, each of which has its own strategic planning facility. In the second area I have worldwide charge of certain institutional and governmental relationships.

A brief word is needed on American Express so that the reader will understand the full degree of its involvement in communications and information in the United States and abroad. Although the company is best known for its credit cards and travelers checks, its biggest business is domestic insurance. It is important also as a commercial and investment banker in fifty countries, excluding the United States. It is engaged in international publishing, catalogue sales, direct mail, and similar services. It is also a partner in a joint venture known as Warner-Amex Cable Communications, which has developed Qube, the technology that provides two-way interactive communication by television.

Two-thirds or more of the 46,000 workers of American Express are employed in some way in information processing and communications. From all this it is understandable why we take continuous sightings on the direction communications is taking and where the company itself is going, a task that falls to me, among others, as chairman of the strategic planning operation.

PUBLIC, PRIVATE, OR QUASI

Let us look first at the role of the PTTs twenty or thirty years hence. Most national telecommunications agencies are government-owned and controlled; the main exception is the United States, where the American Telephone & Telegraph Company is very much government-controlled but considered, somewhat mistakenly in my opinion, to be in the private sector. I believe that there will be a converging of the two types, public and private, into quasi-telecommunications authorities. There are various possibilities.

Already in many countries of the world—Europe, Canada, and Japan, in particular—government participates in public sector communications activities. Think a moment of British Telecom's Viewdata project, Canada's Department of Communications' Telidon, or the French Telematique policy, which has resulted in the electronic yellow pages experiment. Are these public or private sector?

In the United States we see a difference that may be more nominal than real. AT&T is not in what many now consider the truly private sector. First of all, it is nearly 100 percent regulated. Granted, AT&T is privately owned; yet its investment returns as well as its activities are regulated and to that extent it is not quite in the private sector. A part of it might be deregulated in the future, but its basic phone systems seem destined to remain in the regulated sector.

As deregulation suggests, the United States appears to be moving in the opposite direction from that taken in Europe. With the creation of companies like MCI Communications and Southern Pacific Communications Corporation (SPCC), this country is apparently going private. Meanwhile, the publicly owned German Bundes Post, the French PTT, and the British Telecom are expanding their activities into areas that we consider private. Nor is this limited to telecommunications. The significance of all this is of critical interest to large multinational corporations.

These movements are reversible, and if there is a change I believe that the greater likelihood is that Europe will copy us rather than our copying Europe. It is true that much of the European push will be directed by the worldwide economic slowdown, which would tend to promote the growth of the public sector. It is a distinct possibility, however, that Reaganism and Thatcherism could become dominant ethics. Thus there are two broad scenarios: one, the French-German move toward statism and the other, the British-American move in the other direction. The jury is still out in one case, whereas in the other it has not even been addressed.

What difference does all this make? The answer depends on where one is coming from in the political context. I think there is something to the proposition that the private sector is doing a slightly better job than the government in some areas, which could be persuasive. But I would not write off the argument that there is a government interest in making sure that when monopoly exists the public must be protected against predatory actions.

WHERE ARE THE MULTINATIONALS GOING?

Where will multinational corporations fit into the long-term communications future?

The multinational corporation is not solely a phenomenon of the United States, although for a short time after World War II it did seem to be characteristically an American institution. Among the twenty-five largest banks in the world only three or four are now American, whereas twenty years ago most of them were. In 1981 European multinationals of all kinds are firmly entrenched in the ranks of the world's largest enterprises, as are the Japanese.

What is not taken into account is that the fastest growing multinationals in the world are neither European nor American. This award goes to the state-owned petroleum companies of OPEC: Petromin in Saudi Arabia, Pertamina in Indonesia, Pemex in Mexico, and Petrobras in Brazil. These young giants are becoming more than just oil distribution or pumping companies; they are actually going into many businesses around the world. The fast growth of this corporate segment is a phenomenon that will continue into the twenty-first century. We look forward to a twenty-first century with both public and private sector multinationals doing business in communications.

Another huge group of multinational corporations that no one identifies as MNCs consists of the Eastern Bloc trading companies. Today the Moscow Narodny Bank and Soviet trading corporations are important MNCs; they are massive and worldwide. Nor should one forget the Japanese trading companies.

Students of the MNCs are tending to focus on the smaller part of the doughnut. I submit that MNCs will be very much a part of the scene in various forms. They will include state-owned entities like those that exist in the Eastern Bloc and the oil exporting countries. They will also include United States-style multinational corporations essentially in the same form in which they exist today; they won't disappear into a black hole as some would like.

Another phenonemon is in the offing. It is highly likely that large joint ventures of American, European, and Asian companies will emerge, largely for the pooling of risks and capital, which, in turn, will produce giant and competing communications conglomerates. These truly international MNCs will operate in regulated and unregulated environments to bring about major economies of scale. They will also service individual and business consumers quite efficiently. It is difficult to foresee parallel governmental regulatory apparatus developing to regulate them; instead, intense competition may achieve the same purpose.

WHERE TECHNOLOGY MAY TAKE US

I am not a technical person and I tend to see technological changes in a market context. But this outlook yields a certain advantage, I think, in relating what we can do technically with what people will buy. It must be viewed not only from the perspective of wholly new technologies, such as laser applications, but perhaps from the convergence of existing technologies.

We have new connections, we are told. An example is a television set and a data bank in a computer. There are fancy ways to describe this combination, but I prefer to keep it simple. A typewriter connected to a television set becomes a "word-processing system." Hooking a television set to the phone gives us "videoconferencing." What I am getting at is this: Technologies themselves, whether new or made more sophisticated by new connections, mean little in the abstract. We must take into account human considerations and factors like inflation, the perception of reduced real incomes, and future changes in consumer psychology that influence demand.

Let us pause for a moment on the matter of consumer psychology to locate some of the traps.

When you ask people what they want in the future, you get the kinds of answers that American Express studies have turned up. It seems that people want these things: much certainty but some surprise; more convenience but also some adventure; a lot of attention but some privacy as well; more intimacy but also urbanity—they want to be in the swim. And they want complete as well as special information on an ever-increasing number of subjects.

Putting it another way, they want everything: both travel and leisure and food and wine. None is inconsistent. The question is, how do people actually behave when confronted with a new choice of services created by technology?

As experience shows, it is not clear that the mere availability of technology means automatic acceptance of what that technology can do. A brief example is electronic funds transfer. Five to ten years ago many were forecasting a cashless society by 1980. This has not happened; there is more paper by way of checks in our financial system than ever before. People have accepted some aspects of electronic funds transfer but hardly all of them; and many people do not want anything to do with it at all.

We must heed this cautionary note in thinking about the acceptance of the advanced communications technology that is available now. I think it will *gradually* be integrated in a way that will change the conduct of businesses as well as the nature of their products.

The stimulus for change will be a renewed and intense competition for the consumer's business among companies that develop more sophisticated products. In this competition the traditional distinctions between insurance companies, banks, and other participants in the financial industry will become blurred, a process that has already begun. Insurance companies are experimenting with sophisticated financial planning. One property and casualty insurance company, for instance, recently initiated a program in which agents' commissions are electronically deposited in a money-market fund controlled by the company. The agents may withdraw funds by telephone or check. Brokerage houses, on another front, are offering bank-card services and access to free credit balances, whereas a growing number of banks are offering their customers telephone bill-paying. Recently, a major insurance company bought a big Wall Street investment banking and brokerage house, even as the company I represent agreed to buy another.

Soon the consumer will be able to bank by mail, automatic terminal, or cable television, all with equal ease—if the consumer so desires. Around-the-clock banking will then become available in the comfort of the home, if the consumer so desires. Once again, however, we do not know how many people really want these conveniences.

NEW PLAYERS IN THE GAME

The emergence of the new financial services will be greatly facilitated by a steady loss in a distinction between information processing and communications. The merging of these two technologies will make it possible for financial service institutions to provide new and improved products, and this same reliance may encourage the emergence of a new breed of competitor. It is advisable, therefore, for people in my industry to take a close look outside the current boundaries of finance to determine just what organizations, because of their proficiency in the underlying information processing services and communications technology, could become the new competitors within the financial industry. The point is that the new technology will bring new players.

The new technology can do this to any industry, but because we started with financial services we will stick with them. Who, one wonders, will be the big players in the financial industry ten or twenty-five years from now? The technology is around, and it is a little hard to tell who the players are going to be. Take the Exxon tiger that rolls to the offices at night: What underlying technology investment is there that might be refocused to new markets?

It would certainly be unwise not to consider where AT&T might be heading, not just in the next few years but over the next several decades. For my part, I must confess that the company's recent, voluminous filings with the Federal Communications Commission are puzzling to me. Without examining the merits of the various major issues concerning AT&T, I simply would classify that company as a potential major new player in many industries.

In conclusion, here are four working assumptions about the twenty-first century that are worth considering from my vantage point in the early 1980s:

First, most of our commercial and governmental institutions will survive the twentieth century.

Second, the multinational corporation in its various forms, including the state-owned entities that now exist in the Eastern Bloc and oil exporting countries, will be around; so will the U.S.-European style of multinational corporation.

Third, the distinction between "domestic" and "international" will have largely disappeared for many practical purposes, although we will have sovereign nation-states roughly similar to what we have at present.

Fourth, the nature of the business enterprise and the nature of communications will be so interrelated that they will be largely the same subject.

15

PREPARING FOR
THE NEW ERA
IN COMMUNICATIONS

A Japanese Perspective

KIMIO UNO

Almost every nation today is involved one way or another in developing information and communications networks based on the new technology. A determined effort is being made in many cases to push the new trends forward in a quest for deeper knowledge, better services, improved efficiency, and increased reliability. Cumulatively, the countless changes and improvements now taking place or being planned will bring about a world quite different from the one we know today.

The question raised in this chapter is the degree to which society is prepared to deal with the many issues that are raised by this extraordinary worldwide development. My emphasis is on what is happening in maturing societies, with examples drawn largely from the Japanese experience with computers, telecommunications, and information systems.

The communications age has been received with a hearty welcome by some. At the same time warning voices are heard. There is concern that multinational corporations engaged in communications and information processing may come to exert excessive power in the future. Another concern is that the collection of data, coupled with increased capability for storing and analyzing them, may result in an invasion of privacy. Underlying this is a fear that society is heading toward excessive planning and control. I concede the reality of these fears but believe that if society prepares itself adequately the implicit dangers can be warded off.

On one point everyone is agreed: Communication now plays a vital role in today's society and will play an even more vital one in the future. But what do we mean precisely by communication? Exactly what have we in mind when we say that the communications age has arrived? What is the implication for a mature society in which a high level of material productivity has already been achieved? In answering these questions, let us first clarify the basic terms we use—namely, information, information processing, and communication—and examine how they relate to one another.

WHAT DO WE MEAN BY COMMUNICATION?

Information in its broadest sense can be classified into two types according to whether or not it diminishes uncertainty. The first type, which is the subject of this chapter, can be classified further as program or parameter information. Program information refers to the systematic *way of thinking* that is required to solve a given problem; computer programs are an example. Program information is essential in dealing with nonroutine activities that tend to increase as the society becomes more complex. Parameter information refers to the *data* required to use the program information to solve a given problem.

The other broad type of information can be categorized roughly as entertainment. To economists it is similar to other goods and services in its economic and social functions. Movies, television programs, and novels are obvious examples of this type of information.[1]

Information processing refers to the act of gathering, storing, analyzing, and presenting information. Communication is a process of transferring information from place to place or from one transaction to another. One important characteristic of the present-day development is that communication and information systems are being merged into an integrated network. Telecommunications and computers are only small parts of it.

It is important to distinguish between the concepts of "flow" and "stock" as applied to communication and information. Because communication is a process for transferring information, it therefore involves the concept of flow. Information, on the other hand, refers to a stock of knowledge, and we specifically say information *gathering* when we refer to flow aspects. According to this classification, "communications system" refers to those functions concerned with flow, whereas "information system" refers to the storing of parameter and program information, which includes access to this information and its distribution. It should be stressed that perhaps the greatest advantage brought about by the computer is its capacity to store information. Information in the past has been neglected because of the difficulty of retrieval due to the limitation of human memory. Data bases overcome the limitation of human capacity; the electronic file, which stores documents and tables in their original form, will increase the capacity to store information.

It is also essential to identify the agents who actually carry out the process of communication. Communication not only takes place between geographical points but also between various transactors. These include businesses, central and local governments, nonprofit institutions, and households, not to mention multinational corporations whose role is trebly important. They are producers of information and hardware, operators of communications channels, and also the organizations within which program information is shared, regardless of national boundaries. This often makes a multinational the key transactor in the dispersion of technological innovation.

We must distinguish further between universal versus local information. Information may be available only within a limited area or a small segment of society and is often limited to a single organization. The increased sharing of common scientific knowledge throughout the world and the technical progress that has made a worldwide communications network a reality have created the possibility of a universal stock of information.

There are many ways of distributing these scarce resources. It can be done through the market mechanism, equity arrangements, or clubs that transactors join voluntarily. But no matter what solution is attempted, there will be difficult problems due to the technical nature of information and the complexities that arise from its public goods aspect, economies of scale, and decreasing marginal cost. These intricacies prevent the market mechanism from providing the complete answer to the distribution problem. What is more, *information will remain the ultimate scarce factor in a mature society.* In view of the fact that any scarce factor tends to dominate the society, the availability of information will have political implications, domestically and internationally.

Finally, it is important to have an understanding of the substitution effects that result from the spread of information and communications networks. We want to know what they are replacing and what the consequences will be when the new technology is used in combination with older goods and services. Some may be replaced by new technology and social organization as new ones are created. Thus information and communication have repercussions on the industrial structure and on the levels and patterns of employment. They also affect regional distribution of various activities; some factors invite concentration of activities, others foster dispersion. Such forces are at work not only within countries but also across national boundaries. Transportation of raw materials, final products, and people has been an important economic activity. But as societies mature and move into the information age communications channels in many cases will replace transportation. Thus there is evidence of a basic shift in the social infrastructure from the transportation to the information system.

What will be the most efficient method of communication—point to point, point to mass, or mass to point? Here we focus on the new technologies for data collection and distribution, which include communications satellites, increased use of computers with all sorts of terminals (word processors, facsimile, and copy machines), and the more conventional means of communication—the telephone, telex, and printing. It is noteworthy that mass-point communication is now becoming feasible both technologically and economically. Examples are market transactions and voting. Purchases by individuals are aggregated as market demand to determine the resource allocation of society; individual casting of a vote is expressed as a choice among candidates, which in turn represents one platform or another. Recent technology has made both types of polling possible by the interactive cable television systems that are discussed in other chapters of this book.

COMPUTERS IN HOMES AND OFFICES

In examining the impact of the communications revolution on a mature society let us begin with the Japanese household. Ten years ago the telephone was just a telephone that allowed people to communicate verbally. The telephone line is now used to give access to computers. We are able to reserve train tickets by talking directly on our house phone to an on-line computer. We then have access to another computer that carries out calculations which are transmitted back to our phone by synthetic voice.

Thanks to on-line networks among banks, one can automatically draw from one's account to pay for electricity, gas, telephone, water, insurance, tuition, mortgage payment, and credit card charges, among other things. Automatic cash dispensers, also widely used, overcome the barriers among different banks. Medical information, like that contained in an electrocardiogram, is transmitted from local clinics to key hospitals for evaluation, thus eliminating the need for patients to visit specialists. Some educational institutions now communicate with students in their homes by facsimile.

In intercontinental communication the link is usually by satellite. We can call foreign countries directly by dialing the country code, the area code within that country, and finally the subscriber's own number. Domestically, we have answering phones that not only record the voices of callers but

High volumes of customer transactions are handled by computer at this bank in Miyazaki City, Japan. Courtesy of International Business Machines Corporation

enable us to get a playback of the calls from remote areas by dialing our own numbers, or we can have calls transferred automatically to whatever number we want.

In the early 1970s electronic calculators became available at a price the individual consumer could afford. Now the personal computer is becoming popular. Besides carrying out the kind of operation that once required a fairly large computer to perform, it serves as an information file for keeping mailing lists, budget records, and so forth. It can be used as a teaching device to help build vocabulary or to teach mathematical skills, and it is capable of serving as an intelligent terminal that can be connected to a larger machine by public telephone lines.

Television sets are now frequently used in combination with videotape recorders, which both record and edit. Television programs are becoming more specialized, especially those produced for cable television. Multilingual broadcasting is a reality; when watching a movie, one can choose to hear the original voice in a foreign language or a dubbed-in voice in one's own language. The TV screen is being used for TV games or as a monitor for a personal computer. It can also be used for educational purposes through the playback of videotapes.

The business sector is both user and producer of information and communications equipment. As a user, business has moved from the age of the punch card and electronic data-processing system into the age of the on-line network and office information system. The capability of the computer has increased dramatically, the most important advance in ten years being network formation. Computers are now accessible on many terminals, and the construction of data bases has become a major preoccupation that permits multipurpose use by multiple users.

Up to this point the development of the computer can be characterized as computer-room oriented, but now we are witnessing a shift toward user orientation. Although large general-purpose computers are still located in computer rooms, office computers, word processors, and facsimiles can be found on desk tops. Moreover, we are experiencing a shift from numeric to nonnumeric information, which includes graphics and documents.

The introduction of these technologies is drastically changing the business information system in Japan, for management is now able to direct and monitor performance more effectively to obtain instantaneous response.

The new technologies are also adding new dimensions to the competitive ability of Japanese corporations. Particularly important is the development of data bases. Today's managers depend on a bibliographical data base of scientific articles to gather information on the new trends in research and development and marketing. They depend on a statistical data base, frequently located thousands of miles away, to analyze the new trends in the economy of a foreign country or their own, or of particular regions or segments of the market. However, even with the new technology, offices and conferences will not be eliminated because their main function lies in developing and sharing program information.

COMPUTERIZATION OF PRODUCTION AND TRANSPORTATION

In Japanese factories we have industrial robots designed to carry out jobs that are hazardous or require physical power. We have numerically controlled machines and computer-controlled warehouses. To embody these technologies a robot-producing factory has been constructed, operated by industrial robots and numerically controlled metalworking machines and linked to a computer-controlled warehouse by automatic transporters.[2] The use of robots is spreading from welding and painting to various assembly processes. In one case 60 percent of the entire assembly process will be carried out by robots within five years, thus reducing the number of assembly-line workers by as much as 70 percent. They are to be retrained as software engineers.[3]

Currently, there are 75,000 industrial robots in Japan, or 80 percent of the world's total. This number is not inconsiderable when compared with the total labor force, especially when one considers that these machines can be operated twenty-four hours a day.[4] The human labor force released by robots from dangerous and dirty jobs will be able to concentrate on more sophisticated and more creative work. The gain in productivity will result in economic growth and higher income. In societies in which workers refuse to shift to other jobs, however, the result may be unemployment in some job categories.

Subway trains in Japan are now operated automatically by computer. The dual-mode operation of buses is in the experimental stage. These vehicles can travel on special tracks, guided by electronic signals and responding automatically to the calls of riders, or they can be operated manually in congested areas. Traffic signals are computer-coordinated to produce the best traffic pattern. In Tokyo 4900 automatic detectors monitor traffic conditions at various points and feed information into a central system that links signals at 4600 intersections. In the postal service, sorters read the hand-written zip code electronically.

Computer-controlled telemeters are finding wide application monitoring the quality of the air to prevent pollution and the microscopic movement of the seabed to predict major earthquakes. Computerized typesetting (CTS) was introduced in the mid-1960s and is widely used in newspaper and other printing. It was intended to rationalize the printing process while discontinuing the use of lead, a major cause of pollution.

The production of information and communications equipment is the most rapidly expanding segment of the economy. In recent years there have been two major developments. One is the integration of the computer into communications networks, which is bringing about the exchange of patents between the producers of computers and communications equipment. Manufacturers in the United States have more than 80 percent of the world market: The computer industry is virtually nonexistent beyond the United States, Japan, France, and the United Kingdom. Some change in the dominance of American producers may be brought about as the technological gap narrows; moreover, in many countries the communications sector is owned directly by the government or operated by quasi-government public corporations which could strengthen the bargaining power of the users.

The second development is the eclipse of the computer mainframe as the fastest growing segment of the market. In 1979 the production of computers in Japan reached 458 billion yen ($1 equals about 210 yen), but an additional 531 billion yen in peripheral equipment was produced in the same year. While the output of computers has been growing at a rate of 19 percent a year, production of VTRs (videotape recorders) over the last several years has been increasing at an annual rate of 47 percent. Meanwhile, industrial robots and other computerized industrial machinery have been gaining at a rate of 47 percent. The rate of gain for facsimile, word processing, and other office information equipment has been about 40 percent. The size of the market for this equipment in 1981 is expected to reach 127 billion yen, 100 billion yen, and 260 billion yen, respectively. All this is adding new dimensions to the communications and information industry in the 1980s.

Economic activity is becoming knowledge-intensive. Marketing, research and development, quality control, purchasing, storing, distribution, finance, and insurance are gaining importance in relation to production. With the expansion of the service sector, which includes medical care, education, and various community services, the Japanese economy will require increased input by professional and technical personnel.[5] Also needed is the integration of data that are being accumulated in all sectors of society.

GOVERNMENT'S ROLE

The increased role of information processing and communications in a mature society must be shared by the public sector. As the supplier of public services, the government will have to improve productivity, which is currently measured by the input of resources in the absence of proper methods to measure output of government services.

As the economy matures, extra-market decisions tend to become important. In certain areas automatic adjustment by market mechanisms cannot be relied on, and deliberate information gathering and processing are required. Defense, highways, airports, and education are among the services that fall into this category. There are cases in which the externalities cannot be dealt with properly by the market mechanism: Environmental pollution is an example. There are other cases of decreasing marginal cost in industries in which public bodies of some kind must be established: for example, telephones, electricity, and perhaps railways. There are cases in which the society agrees that a certain minimum level of subsistence must be maintained regardless of the contribution to productivity, as in education and welfare. There are others in which certain macroeconomic policies are required to make adjustments between aggregate demand and supply or between production and distribution to avoid inflation or unemployment.

The performance of society from an economic, social, and perhaps political perspective must be monitored by statistical data collection. Various government branches, as well as nonprofit organizations in such areas as education, Medicare, and pensions, have to deal with common problems, although from slightly different angles; it is therefore imperative that they have access to common data sets. Despite the need for an integrated approach to common problems, most government organizations are organized vertically rather than horizontally to reflect the changing social goals.[6]

THE ROLE OF UNIVERSITIES

A major question concerns research institutions and universities: Are they prepared to handle the new role into which they are being thrust by the communications age? These are the places where computer and communications specialists, economists, urban planners, sociologists, psychologists, and political scientists meet to prepare society for the potentials of the new technology. To be responsive to the requirements of society we need to break down the barriers between different disciplines. Actually, if we were to look within each discipline, we would realize that different sciences are using surprisingly similar analytical tools. These disciplines have many things in common. Mathematics, statistics, and computer science, plus English as the working language, are common means of communication the world over.

Interesting interdisciplinary developments are taking place at the Institute of Socio-Economic Planning of the University of Tsukuba. The Institute is one of twenty or so similar organizations at the university. These are comparable to traditional academic departments except that they cross over the usual lines. The Institute encompasses the disciplines of economics, managerial science, and urban and regional studies and is developing three experimental programs that bear on communications and information processing. The common languages are economics, econometrics, and computer science.

The first of these programs, the Social Behavior Experiment System (SOBES), consists basically of a set of response analyzers linked to a computer. About forty people, each equipped with a response analyzer on his or her desk, gather in a room. The devices, which look more or less like electronic calculators, are flanked by magnified videotape and slide projector screens that are used to convey visual, verbal, and numerical information. Viewers may be asked to evaluate a regional development plan or to respond to a political candidate's opinions.

In opinion polling it usually takes time to compile the data, but the SOBES method provides instant compilation by various criteria, such as age, occupation, sex, and experience. The result is reported to the viewers instantly, which allows a second round of response. A typical phenomenon in polling is that the lack of key information on an issue tends to make opinions diverge, whereas with sufficient information opinions will tend to converge. SOBES helps us to evaluate the distribution of opinions within the group and to determine whether and why significant divergences of opinion exist.

This system can also be used to explore group dynamics. Several rooms are equipped with graphic terminals that permit conversation by computer among groups. One group may take the position of the business community, the other the position of consumers, or one group may take the position of the seller, the other of the buyer. The patterns of negotiation and convergence or divergence of opinion are recorded and analyzed by computer. This system is an example of what can be called "opinion technology." It makes possible an interplay between man and machine, and it is dynamic because information is fed into the system during the discussion.

Wise use of these devices will help us to overcome the flaws present in the political process, particularly in the voting system. Although it is the basis of representative democracy, voting clearly has certain deficiencies—among them the intervals that occur between choices of alternative policy packages. Yet direct participation of all members in each decision is not always desirable because of the cost and time involved. The result is that the voice of the silent majority is not necessarily reflected in policy decisions. The development of systems such as SOBES can therefore help to bring about greater participation in the democratic process.

The Institute's second program, called the Regional Space Simulator, is essentially a system concerned with input, analysis, and output, in which information is accumulated in different forms and must be integrated. Data gathered from various types of census—population, housing, and manufacturing—are too voluminous to be handled in a conventional way. They provide more valuable information when used in conjunction with maps that show physical contours, land use, zoning, roads, and waterways. Analog information contained on maps or satellite pictures is read into the system by a digitizer or color graphic scanner; grid data on population, production, and housing are also read in. Information on the location of public facilities like schools, hospitals, and community centers can then be integrated. Finally, the data can be analyzed and the results presented on a screen or plotted on paper. This system is used in drawing up regional plans for urban renewal, new residential development, and environmental assessment, or for planning escape routes in case of flood or earthquake.

The Socioeconomic Simulator and Data Base, the Institute's third project, is intended to cover five broad areas: economic, regional, managerial, social, and political. The simulator allows editing, presentation, and analysis of the data, plus the construction and operation of simulation models.

In the field of economics Japan has for a long time compiled standard national accounts that are comprised of national income statistics, input-output tables, statistics on the flow of funds, and various ongoing statistical tabulations of physical as well as financial stocks. There has been an emphasis more recently on other kinds of data that reflect economic transactions involving households, local and central governments, and other nonprofit organizations such as educational institutions, hospitals, and pension funds. As social concerns in Japan shift from macroeconomic to quality-of-life issues, it is increasingly difficult to cope with those problems in terms of macroeconomic data. A list of valid social goals includes the improvement of learning, employment, quality of working life, leisure, and personal economic conditions; others would involve the social environment and participation in social activity.[7] In developing industrial policy aimed at promoting certain key sectors of the economy, it is also necessary to have data on the micro level.

Economists, political scientists, psychologists, regional planners, administrators, and business managers now all use the same sets of data and strikingly similar analytical tools. There is much to be done in developing integrated data bases and promoting cross-disciplinary analysis. But this poses some serious problems, one of which is the shared use of statistical data that require careful development of a data-base management system. This has been done to a certain degree with business data, but the problem is more complex with a data base for research purposes.

The socioeconomic simulator is designed to assist in the manipulation of policy variables in the model from several graphic terminals, thus making it possible to base policy debates on a single objective model; for example, representatives from the finance ministry, central bank, labor unions, and consumer groups can be given terminals. One person may recommend a tax cut, whereas another may propose increased government spending. Each policy recommendation may be compared graphically by use of the simulation model.

The Institute of Socio-Economic Planning has access to statistical data bases developed by other universities and research institutions, some of which are located overseas. Official statistics are also being made available in the form of computer tapes; some day these data will be accessed directly from remote terminals, and there are plans to facilitate computer conferencing by using facsimile to transmit graphic information efficiently by telephone lines.

THE PROBLEM OF PRIVACY

Thanks to technological progress in information and communications, it has become possible to integrate pieces of information. This raises the problem of privacy. We have been graded in schools. We have been employed and have filed income tax returns. We have opened bank accounts and obtained credit cards. We have obtained driver's licenses, taken out insurance, and been hospitalized. We have responded to the census. We have cast votes for one candidate or another. But now we are faced with a new set of problems.

When used and examined in conjunction with other pieces of data accumulated over a period of time, a single piece can produce more information than was originally intended. That this might happen without our consent understandably produces anxiety. Important though it may be to assess or to base public policies on objective data, nevertheless we do not want to endanger our privacy.

Today's policy decisions need detailed data on income distribution by occupation, age bracket, region, and other criteria. Business decisions likewise require detailed data on segments of the market. In the academic field, if microdata are kept and made accessible, economists, sociologists, psychologists, and regional specialists can cooperate across the boundaries of disciplines. Aggregate data, although important as the control total, are considered of limited use. It is also true that aggregate data are more useful when based on microdata accumulated in consistent series.[8]

Does this inevitably endanger privacy? Not necessarily. We should be able to compile new data sets from existing microdata sets based on individual samples without revealing individual identity. Limiting access to processed data is quite possible. Analytically, this is often sufficient because we are usually interested in the characteristics of a subgroup of the sample, not in individual identity. Provisionally, we should limit access to individual records in the information-processing system in which they originate.[9]

More detailed safeguards can be developed as society has more experience with far-reaching information networks. We are still at an early stage in the process of readying ourselves for the new era of communications.

CONCERN ABOUT MULTINATIONALS

We come to a final area of concern: the danger of monopoly. There is certain justification for fearing that some vast monopoly power may emerge in our nations or that multinational corporations may monopolize international communications channels. It is true that the information and communications system has to be standardized to enhance efficiency and interchangeability. The nation-states used to be a good proxy in delegating business planning decisions. After removing the barriers of international trade and investment, and to a lesser degree those imposed by the labor force, national boundaries are becoming less important economically, if not politically. Consumers and businesses alike are keeping watchful eyes on information in remote countries. We have multinational producers, multinational distributors, and multinational consumers. No wonder the communications system has become multinational to meet these demands.

Looked at from the supply side, this seems to be one of the areas in which economies of scale obtain. The particularly huge expenditure in research and development required for information processing and communications works as a barrier against entry. People are therefore concerned that there may not be many firms left, despite the size of the world market. Not many countries will be able to afford even one information and communications enterprise.

In addition, there is the problem referred to earlier of an industry with decreasing marginal costs: As the size of the operation expands, the additional unit can be produced at a lower cost. In such cases the competitive market will fail to supply the goods in question because the usual way of determining price (by setting it to equal marginal cost) would necessarily involve loss to the supplier. If information and communications are actually such a case, then the competitive market would not prevail.

It is argued that once multinationals have dominated the market they will almost certainly send their earnings to a location that will allow them maximum profit, taking into consideration such factors as taxes. Technical progress could be hampered because the MNCs might cut their expenditures on research and development, continue to use existing equipment, and delay the introduction of new equipment in any given country. It is therefore argued that host countries may derive no benefit from operations within their borders. Furthermore, information and communication are such vital facets of a modern society that the very ownership of these channels by organizations beyond national control creates fear in regard to economic and political independence.

This applies to producers of information and communications equipment, operators of communications channels, compilers of bibliographical and statistical data bases, and distributors of information. Smaller units tend to be integrated into a larger organization from the standpoint of supply and demand, and it is to the users' benefit when they are integrated. A data base becomes more comprehensive as communications channels and equipment become more compatible and efficient. The users of data bases gain access to various other data bases from a single distributor.

At the same time we should realize that there are opposing trends, some only just developing. In the field of hardware technological progress seems to have brought about a reversal in the movement toward larger and more centralized systems. Decreasing cost has made intelligent terminals a reality, and we can observe trends toward decentralized systems. Improved communications channels are making them technically possible and economically affordable. Technology is becoming more diversified as widespread and specialized applications make room for smaller firms to enter the market. Here is an area in which new entry is possible without huge initial investment. Much depends on the efficient organization of specialists and knowledge workers. Data-base makers and their distributors are becoming more independent of one another. There is a strong trend toward local suppliers filling specialized needs rather than huge general ones.

From an examination of all the available evidence we cannot conclude that information and communication bring about monopoly and centralization. Although it is premature to say that they are inherently decentralized, recent evidence seems to suggest decentralizing trends.

In general we see that information processing and communications technology are not electronic monsters about to dominate our society. On the contrary, they increase the human role in society. We are now able to establish feedback channels in places where it was formerly impossible.[10] The rising production of information and communications equipment, the increasing input of knowledge workers, and the growing stock of knowledge are important. But the true significance seems to lie in the fact that, thanks to these new developments, the way our society is organized is being changed drastically. This opens new horizons for all of us, economically, socially, and politically.

Notes

1. Yukio Noguchi, *Johoono Keizai Riron* (The Economic Theory of Information), Toyo Keizai Shinposha, Tokyo, 1977.
2. *Japan Economic Journal,* January 16, 1981.
3. *Japan Economic Journal,* March 3, 1981.
4. *Japan Economic Journal,* January 12, 1981.
5. See Kimio Uno, "The Communication Sector in Japan and Its Role in Economic Development," Workshop on the Economics of Communication, the East-West Communication Institute, June 1980.
6. See, for example, Edward Schneier, "The Intelligence of Congress: Information and Public-Policy Patterns," *The Annals of the American Academy,* March 1970.
7. Organization for Economic Cooperation and Development, *1976 Progress Report on Phase II,* Paris, 1977.
8. For recent discussion of the needs and methodology of integrated microdata, see, for example, United Nations Economic and Social Council, *Methods of Collecting, Organizing, and Retrieving Social Statistics to Achieve Integration,* E/CN.3/516, June 5, 1978; *Role of Macro-Data and Micro-Data Structures in the Integration of Demographic, Social and Economic Statistics,* E/CN.3/552, July 3, 1980.
9. Ryosuke Hotaka and Hideto Sato, "Database Approach to Statistical Data," Fourth Japan-U.S. Forum on International Issues, Tokyo, 1980; Hideto Sato, "Derivability and Comparability Among Non-Atomic Data," The Seventh International CODATA Conference, Kyoto, 1980.
10. "Voting or not voting for a political party or a candidate, buying or not buying a product, leaving a job or staying, often have the character of bulky 'package deals.' It is only in rare instances that the expression of such choices can be taken as pleas for specific social action. Moreover, people frequently are not sufficiently articulate, informed, or otherwise able to translate dissatisfaction with outcomes into preference for societal action." Organization for Economic Cooperation and Development, *Subjective Elements of Well-Being,* Paris, 1974.

16

THE FREE-FLOW DOCTRINE

Will It Last Into the Twenty-First Century?

HERBERT I. SCHILLER

A central polarity dominates international communications in the last decades of the twentieth century. Transnational business pulls powerfully in one direction. It needs a maximum free flow of information in all aspects of its affairs. Against this is another powerful force exerted by the majority of the world's nations and by their political leaders and power centers. These leaders perceive it to be in the best interests of their societies to put limits on the free flow of information for a variety of reasons—social and cultural, political and economic. As we move closer to the twenty-first century the struggle between these two powerful forces intensifies, for the stakes are very big indeed.

The struggle over information is only one aspect of an even larger contest that involves the vital international flows of all goods, tangible and intangible. United States-owned transnational corporations are equally adamant in their advocacy of unimpeded capital flows and of their right to invest or disinvest wherever they choose. The same companies are just as dedicated to the unobstructed flow of raw materials, certainly of a one-way flow into the United States or wherever they want them stockpiled. Total dedication to the broad principles of free trade may vary according to whether an industry is in the sunrise, or more dynamic, sector or in the sunset sector of older, aging industries which incline more toward protectionism. Such internal differences as these, however, cannot obscure the vast differences in trade and investment policy centering on transnational corporations that have been waged for several decades between the United States and much of the rest of the world, particularly the developing countries.

This chapter attempts to put the growing contest over worldwide information flows into the context of this far larger struggle. In other words, the whole doctrine of the free flow of information, which began as a more limited concept, now intersects with the larger issue of international trade and investment at most points. In the new global context information and organization, technology and production, goods and needs, and capital as money and capital as knowledge cannot easily be separated.

TOTAL DEPENDENCY ON COMMUNICATIONS

There is complete unanimity of view on free flow within the information industries and the information goods and services sector of the U.S. economy. Maximum free flow continues to be the nonnegotiable position of the industries' representatives. Harry B. DeMaio, director of security programs for International Business Machines Corporation, expresses the general sentiment:

> Regardless of the issue, I believe the paramount objective should be the preservation of the free flow of information across and within national borders balanced by considerations for privacy and national security. This must be the single most important objective simply because of the immense economic and political impact of free information flow in today's society.[1]

There are compelling reasons why this view is so strongly held. Any single transnational corporation in itself is a system of world enterprise that binds together perhaps scores of operating facilities in dozens of countries. This necessitates a dense network of reliable, rapid, uninterrupted, and unrestricted informational circuits. The book value of the overseas plant and investment in facilities of all U.S. transnational companies now surpasses $200 billion. The annual value of the output of this aggregate investment exceeds the gross national products of all but a few countries in the world; only Japan, West Germany, and the Soviet Union produce more goods than the overseas plant owned by American capital.

The few thousand businesses that engage in transnational operations constitute the core of U.S. enterprise. These are the companies that produce most of the goods and services, employ a large fraction of the work force, account for the bulk of capital investment, and receive the lion's share of the profits. The largest of these corporations derive a substantial and *growing* portion of their revenues and profits, ranging between 25 and 60 percent, from their foreign activities.

The administration and operation of this transnational business system, which has no historical precedent, requires a level of managerial skill made possible only by flexible and massive communication capabilities. The communication networks, which have been organized to accomplish the transnational corporate effort, provide financial and audit reporting, market research and statistics, inventory control and plant management, and research and development. Some of the data handled in the networks are sales orders, product price information, delivery messages, inventory updating, personnel changes and movement, and monthly financial reports.[2]

The combination of the computer and telecommunications has transformed the manner in which corporations do business. Hugh Donaghue, vice president and assistant to the chief executive officer of Control Data Corporation, says:

> Today a company can collect data on its operations and personnel from all parts of the world, and almost instantaneously transmit such data to a single location for storage, processing and further dissemination. This provides the management of these companies with current information for decision making on a scale never before possible. As a result of these technologically based breakthroughs new concepts of management are appearing. *Underlying these concepts is the requirement for the free flow of information between all elements of the corporation and between the corporation and its customers, vendors, bankers, etc.*[3]

This makes a further striking point. Not only are information flows crucial to the daily transactions of global business, but they are also crucial to the growth of the corporation and the whole transnational sector in those scores of countries that serve as hosts.

Transnational corporations by nature and structure are expansive. Profit is the goal and increased sales lead to it. Expanding markets consequently are essential.

In developed market economies the effective and therefore preferred way of extending a market is by advertising in the mass media. These imperatives are operative in the international as well as the domestic sphere.

Because of its almost total reliance on mass communication channels and information flows in developing markets, the transnational business system has been active in transmitting its marketing messages. In 1980 *Advertising Age* reported that worldwide advertising was estimated at $110 billion, of which half was spent inside the United States. The magazine also estimated that by the twenty first century global advertising expenditures will have amounted to $780 billion.

"The rest of the world is rapidly emulating many of our advertising practices," writes Robert J. Coen, senior vice president and director of forecast and worldwide liaison of McCann-Erickson, "and by the year 2000 these will be the norm in a number of other countries around the world." The active role of U.S. transnational business and transnational advertising agencies in this development receives indirect mention. Coen notes that "many of the most well known U.S. brands are marketed by a division of a large multinational corporation. The advertising strategies now being employed successfully in the U.S. will inevitably spread around the world as media availabilities and consumer purchasing power grows."[4]

The U.S. transnational corporate system is thus totally dependent on the full range of communications technologies and information flows, including access to the media. Without all this it could not carry on its daily operations nor extend its sphere of activity and create its legitimacy wherever it is installed. This dependency far exceeds the earlier need of U.S. media and media products to find overseas markets, thereby bringing about a totally new situation in the world.

In the years immediately after World War II the press organizations and media corporations were the active proponents of the "free flow of information" doctrine on behalf of their own interests, defending this in terms of general social well-being. Although this remains an important element and source of support for the free-flow position today, it is subsumed by the general need of the U.S. transnational corporate system for an unrestricted flow of information. This new situation is explained by Philip H. Power, member of the United States delegation to UNESCO and owner and chairman of Suburban Communications Corporation, which publishes forty-two community newspapers in Michigan and Ohio:

The stakes in the coming battle go far beyond editors and publishers, who, so far, have been the only ones directly involved. They extend to the great computer and information hardware companies whose foreign sales of billions of dollars are at stake; to the TV networks and movie makers whose entertainment products range the globe; to the airlines and banks and financial institutions whose need for computer-to-computer data literally defines their business; to the multi-million-dollar international advertising industry. . . .[5]

All of this led Anthony Smith, director of the British Film Institute, to make this appraisal:

Oddly enough, the field of informatics, which represents the developed world at its most inexorably powerful, is also one in which the same countries are very vulnerable. For if the transformation of Western industrial economies or information economies is to succeed, they must eventually enjoy the benefits of a free flow of data.[6]

Hugh Donaghue of Control Data says much the same thing: "The basis for new management is a growing dependency on the free flow of information, and consequently, a growing vulnerability if this free flow is restricted or stopped completely."[7]

A WORLD OF OBSTRUCTIONISTS

But why would the free flow be restricted, and who would intervene to do it? In the briefest terms, the miscreants are all those who object to the benefits and privileges that "free flow" confers on U.S. transnational enterprises and to the disadvantages and costs that are perceived to accrue to those nations affected by transnational corporate activity. The obstructionists—and potential obstructionists—are a surprisingly diverse group. For Americans, accustomed as they are to casting communists as the primary instigators of global, national, and community trouble, this anti-free-flow coalition presents a disconcerting pluralism.

An example of those who are outspoken in their concern over the flow of information across national borders is Jan Freese, director-general of the Swedish Data Inspection Board and formerly a judge. He has often said that "it seems to be a paradox, but nevertheless the free flow of information . . . has to be regulated by international agreements in order to be kept free."[8]

Our close Canadian neighbors, with whom we share a 3000-mile open border, also have forceful views on "free flow." Although Canadians are naturally far from unanimous, official statements indicate a wide divergence from the U.S. position. An extraordinary committee was created by the Canadian minister of communications in 1978 to study the implications of telecommunications for Canadian sovereignty. Known as the Clyne Committee, it reported this in 1979: "Telecommunications, as the foundation of the future society, cannot always be left to the vagaries of the market; principles that we might care to assert in other fields, such as totally free competition, may not be applicable in this crucial sphere."

The committee was especially concerned with transborder data flow, data-processing services, and external control of national information in data banks. "Few people in Canada," it stated, "are aware of the implications of what is happening." The committee urged that "protective measures" be taken. The committee concluded its report with some powerful language seldom seen in a government document:

> We see communication as one of the fundamental elements of sovereignty, and we are speaking of the sovereignty of the people of a country. . . . We urge the Government of Canada to take immediate action to alert the people of Canada to the perilous position of their collective sovereignty that has resulted from the new technologies of telecommunications and informatics.[9]

The French have been no less critical and apprehensive of the workings of the free-flow doctrine. A paper on transborder data flow written in 1980 by Alain Madec, a senior French official, received this comment in a U.S. congressional study:

> From the perspective of those concerned with the free flow of information and the liberalization of trade, the tone of the draft is disturbing. While asserting a deep regard for the principle of the free flow of information, the report deftly suggests that the world needs to move away from that principle, stating that: ". . . other complimentary [sic] values cannot be ignored, i.e., the responsibility of sovereign states, the balance of advantage to be derived from mutually profitable trade, and respect for the diversity of peoples and cultures."[10]

This is but one of the innumerable official French comments in recent years which have questioned the free-flow doctrine.

Canadian, Swedish, and French leaders have indicated their concern over U.S. domination in the economic and information fields. As might be expected, Third-World leaders from Latin America, Asia, and Africa express even more critical views. They have reason to do so. Their societies are not competitive with the transnational system in any respect. With few exceptions, they experience the full weight and costs of the transnational companies in their midst.

There are, to be sure, important differences in the views of the hundred-plus bloc of national states on free flow, as there are on most other political, economic, and cultural questions. In more than a dozen high-level international meetings of nonaligned and Third World leaders, the free flow of information as it occurs today has been identified as a mechanism for informational penetration, cultural domination, and economic exploitation.[11] This criticism, which began in Algiers in 1973, was recently reiterated at Yaounde (Cameroon) in July 1980. "The grip of the multinationals on world communication" was cited there as a major obstacle to global informational equality. The conferees urged "a radical change in the relationship of communication to knowledge, money, and power."[12]

CHANGING VIEWS

Despite the almost universal opposition to the prevailing international information structure and to the principle of free flow that undergirds it, some in the United States persist in regarding this opposition as "paranoiac," "obsessive," and reflecting a preoccupation with a "bogeyman" issue.[13] A more realistic assessment has begun to emerge, however.

A December 1980 report of the Committee on Government Operations of the U.S. House of Representatives noted the following:

> Whatever the particular perspective of a country, an increasing number of nations worry that the loss of control over information about internal functions can jeopardize their sovereignty and leave them open to possible disruptions, ranging from uncontrollable technical failures to political sabotage.

The committee recognized that privacy protection, national sovereignty, external control of domestic activities, national security, technology transfer, and "cultural erosion . . . are common concerns not limited to the Third World."[14]

Another indication of changing sentiment comes from Anthony G. Oettinger, the chairman of Harvard University's Program on Information Resources Policy, who recently wrote:

> The resulting conflict between the principles of free flow of information and of national sovereignty is fueled by the fear, expressed by many states, that the United States would use its great technological advantage in this area for political, cultural, or commercial purposes. . . . Consequently, the restrictionist impulses of most of the world [are pitted] against an increasingly isolated American devotion to a principle of free flow of information.[15]

A recent statement by Morris H. Crawford represents another step forward in awareness. Crawford was formerly executive secretary of the U.S. Department of State's interagency task force and public advisory group on international information flow and is now executive director of the Brussels Mandate: An Alliance for World Communication and Information. He acknowledges that "a reweighing of past applications of the free flow doctrine" is long overdue and goes on to say this:

> The doctrine itself arises from constitutional principles that no sensible person seriously questions. But these principles have to be applied realistically in terms of the actual channels through which information flows. Other governments, while continuing to respect our constitutional principles, have shown with startling unanimity and unmistakable clarity that they want to take a fresh look at modern application of these principles. Our isolation has grown in the two years since U.N. debates on direct broadcasting via satellites found the U.S. deserted even by its staunchest allies. . . . U.S. assumptions about free flow need to be reassessed in the context of modern communication technologies and their impact on other members of the interdependent world community. *The U.S. is out of phase in an international reexamination in which we have more at stake than any other country.* Paradoxically, failure to face the reality of changing definitions of free flow tends to strengthen the hands of those who favor stronger controls over all forms of global information exchange.[16]

But, if the free flow of information is a vital support to the global overlordship of American transnational companies, how much room is there for adjustment?

Crawford is among those policymakers and other influential people in the United States who think there is space to maneuver. He urges "a comprehensive theme and unifying purpose" for U.S. policy. We should define the boundaries within which some "international rules for communication and information flows might be acceptable." There should also be "coordination and monitoring of U.S. planning and negotiating activities that affect international communication."[17]

Robert Manning, former editor of *The Atlantic Monthly* and associated with State Department thinking over a long time, also has taken up the question of the concessions that may have to be made to keep the overall system intact. He writes:

> There is still time, it appears, for Western government and communications giants to cure this intolerable imbalance by constructing the new communication networks "in a spirit of real interdependence and bringing the people of the Third World into the manufacture, installation and some of the operations of the new electronic technology."[18]

Manning understands that this could involve "some uncomfortable concessions" on such matters as the distribution of radio frequencies and partnerships with Third World governments. "Stockholders might not care for such arrangements," he adds, "but their managements plainly should be thinking about forestalling the conflicts and calamities that [may be in the offing.]"

There is still no assurance, however, that Congressional policymakers have understood this message. They tend to focus less on concessions and more on overcoming the complexities in international communications policymaking by governmental bodies, as demonstrated by the U.S. House of Representatives Subcommittee on Information and Individual Rights.* In December 1980, following hearings held earlier in the year, the subcommittee proposed legislation to improve executive coordination of the international policy process. It suggested the creation of a Council on International Communication and Information to "develop and implement a uniform, consistent, and comprehensive U.S. policy on international communication and information" and to serve as the President's advisor on these matters.[19]

Improved coordination and consultation in U.S. information policymaking, if attainable, undoubtedly would be helpful to a more skillful presentation of the positions of the United States in international forums. It would not, however, address the central issues posed by a worldwide movement against free flow and the hegemonic information and economic power this confers on the transnational corporate system. No matter how expertly co-

---

*A subcommittee of the U.S. House of Representatives Committee on Government Operations.

ordinated and implemented, policy that does not satisfy these basic international economic, political, and cultural demands can yield trivial gains at most.

Where, then, does this leave the free flow of information, that beleaguered pillar of transnational corporate enterprise?

Admittedly, the challenge to free flow is not powerful or united enough at this time to secure the structural changes in the international and domestic systems that would bring about a genuinely new international economic and information order. An uneasy and unstable situation now exists, though for how long is uncertain. Although the defenders of unalloyed free-flow policy are still numerous and influential in the United States, it seems inevitable that concessions will have to be made in the doctrine, probably sooner rather than later. The opposition is too powerful to be ignored; despite rebukes, it is likely to intensify. The House Committee on Government Operations cautions that "the threat of even greater barriers [to free flow] loom." The committee believes that "the potential that a wide range of barriers may be erected in the Third World continues to increase."[20]

The central question is this: How long can the free flow of information, along with the unrestricted flow of everything else essential to the transnational corporate system, survive the pressure from three-quarters of world humanity who are seeking an increased share in global and local resource allocation? Those other flows vital to the transnational corporate system include natural resources, people, goods, and capital.

Further crucial questions also remain unanswered. Can the intolerable inequities that presently disfigure both domestic and international distribution be maintained? How about the expectations of the transnational corporations for the expenditure of a trillion dollars on advertising in the twenty-first century to extend worldwide the consumption patterns familiar in the United States? Can these actually be realized? Will the television programs, films, and other entertainment produced in a small number of western factories continue to preempt world screens and stages twenty years from now? Will U.S. data banks, plus a few more in Europe and Japan, provide the patterned information on which social, political, and technological decisions will be based in Latin America, Africa, and Asia? Will a few thousand companies continue to use 90 percent of the international communications circuits to relay their orders, queries, and commands to their subsidiaries on all continents? In sum, will "interdependence" continue to be defined as binding relationships between unequals?

As in the past, cultural as well as economic, technological, and scientific developments will be shaped by popular movements. There will be continued struggle against established interests to assert new principles and to organize new criteria that will satisfy human welfare and material existence. Developments also will be influenced by intensifying conflictive relations between privileged groups and sectors, each elbowing the other to protect or advance individual positions. Caught in the midst of these warring factions, the free flow that has supported and facilitated the prevailing transnational business system would appear to be facing a stormy future. Seen from this perspective, the twenty-first century looks overcast.

Notes

1. Harry B. DeMaio, "Transnational Information Flows: A Perspective," in *Data Regulation: European and Third World Realities,* Online, Uxbridge, England, 1978, p. 170.

2. Jean-Pierre Chamoux, "The Network Development in Europe," in *Data Regulation, ibid.,* p. 32.

3. Hugh P. Donaghue, "The Business Community's Stake In Global Communication," paper delivered at the 43rd Annual Meeting, U.S. National Commission for UNESCO, Athens, Georgia, December 12, 1979, p. 5. Emphasis added.

4. Robert J. Coen, "Vast U.S. and Worldwide Ad Expenditures Expected," *Advertising Age,* 1–16 (November 13, 1980).

5. Philip H. Power, "Threat to Ad Freedom?" *Advertising Age,* 44 (December 15, 1980).

6. Anthony Smith, *The Geopolitics of Information,* Oxford, New York, 1980, p. 130.

7. Hugh P. Donaghue, *op. cit.,* p. 5.

8. Jan Freese, "The Present and Future Swedish Data Policy," in *Data Regulation, op. cit.,* p. 81.

9. *Telecommunications and Canada,* Consultative Committee on the Implications of Telecommunications for Canadian Sovereignty (The Clyne Report), Ottawa, Canada, March 1979, pp. 2, 63–64, 75–76.

10. *International Information Flow: Forging A New Framework,* Thirty-Second Report of the Committee on Government Operations, 96th Congress, 2nd Session, House Report No. 96-1535, December 11, 1980, U.S. Government Printing Office, Washington, D.C., p. 33.

11. Herbert I. Schiller, "Decolonization of Information: Efforts Toward A New International Order," *Latin American Perspectives,* **V**(1) (Winter 1978).

12. Intergovernmental Conference on Communication Policies in Africa, UNESCO, July 22–31, 1980, Yaounde (Cameroon).
13. Ithiel de Sola Pool, "Exporting Data—Latest Paranoia," *Telecommunications Policy*, **4,** 314 (December 1980).
14. *International Information Flow, op. cit.,* pp. 19–23.
15. Anthony G. Oettinger, "Information Resources: Knowledge and Power in the 21st Century," *Science*, **209,** 191–198 (July 4, 1980).
16. Morris H. Crawford, "Toward an Information Age Debate," *Chronicle of International Communication*, **1**(2), 3 (December 1980). Emphasis added.
17. *Ibid.*
18. Robert Manning, "Data Is Wealth and Power," review of Anthony Smith's *The Geopolitics of Information,* in *The New York Times Book Review,* December 7, 1980, pp. 15, 38–39.
19. Jake Kirchner, "Revamp of Transborder Data Policy Proposed," *Computerworld*, 15 (December 1980).
20. *International Information Flow, op. cit.,* p. 31.

17

NEW INTERNATIONAL DIRECTIONS
A Nonaligned Viewpoint

KAARLE NORDENSTRENG

The international viewpoint expressed here reflects nonaligned thinking that is to be found in Europe and in much of the Third World. My statement is deliberately provocative and is intended to cast critical light on some fashionable visions of communications in the future. I do not challenge the facts about contemporary developments in communications presented by other contributors to this volume. But beyond their statements lies a certain mystification about the future of communications that has resulted in various ill-founded conceptions which must be examined critically and thoughtfully; one might say that they should be "problematicized." Two overall stereotypes, in particular, merit special critical attention. One is conjured up by the phrase "communications age," the other by "global village," which is Marshall McLuhan's contribution to the debate.

The notion that we are entering or living in a communications age suggests that an advanced industrialism is leading essentially to an "information society" in which the dominant social and economic activities center on the handling of information rather than materials. This view is supported by transformations in the structure of the labor force and in methods of storing and transmitting information in the service of production and management. The idea gains further credence from the pressure that developing countries have exerted in UNESCO and other political forums in demanding that all forms of communication serve their interests better. Even Marxist–Leninist positions lend support to the vision of the communications age by asserting that the struggle between ideologies intensifies under conditions of advanced capitalism and socialism.

Yet the vision is as ill-founded and misleading as Daniel Bell's theory of postindustrial society or its variant developed by Zbigniew Brzezinski, the "technetronic era." The trouble with these theories is that social development in technocratic terms concentrates on methods used in production and management, on the lifestyles of people, and on the formal aspects of running society. Absent is a truly historical dimension, which alone can yield an adequate understanding of man's emergence and the development of civilization. Even more significant is the absence of an analysis based on social classes; the class aspect is looked on as an obsolete notion abolished by the "post" stage of industrialism. A profound examination of the social foundations of the information society is very much needed.

A critical approach to these popular theories reveals that they contain more ideological manipulation than social science. One does not have to search far to expose the fallacy of the communications age. It is evident that information has had a crucial role in the history of man and his productive activities, beginning with the Stone Age. Nor are information and communication isolated social phenomena but rather integral parts of others more fundamental. Finally, communication embodies conflicting social interests with the consequence that no piece of social information can ever be completely neutral, independent, or free of the social forces that operate within and between societies.

This kind of ABC is naturally understood by those coming from developing countries in which the struggle for survival is an everyday reality. Thus it is only logical that a movement has been organized in the nonaligned countries to declare a new international information order parallel to the program for a new international economic order. This is difficult for Western thinking to grasp, although there are signs that the enthusiasm for new communications technologies is being tempered by critical reasoning. Elie Abel's chapter offers an example of this more balanced and critical approach. Douglas Cater's recent article on "the survival of human values" offers another example of reasonable Western thinking.[1] Likewise, a sensible call from France has been voiced by François Régis Hutin, who makes the point that "informatics" is "becoming the holy grail of the West . . . the panacea for unemployment and trade deficits, for melancholy and boredom." He has this advice to offer: "Stop. The bandwagon is rolling too fast. The enthusiasm is too forced, too much."[2]

McLuhan's vision of the global village, as recently elaborated by Ithiel de Sola Pool and others, suggests that technological development and economic integration will demolish the nation-state as a frame of social order. This is to be replaced by a more or less universal marketplace of production and consumption, facilitated by the rapid transfer of information and material goods and unimpeded by such "anachronisms" as national frontiers. The global village plus the communications age would therefore produce a paradise in which there is a free flow of information on a planetary scale.

Once again, certain aspects of contemporary development, which include increased trade and other economic exchanges between various parts of the world, more direct contacts among distant societies by mass tourism, and virtually instantaneous communication around the globe, seem to support this vision. But this is merely another false view of reality. National frontiers are not necessarily withering away, nor are societies inevitably being pushed by technology into a homogeneous "United States of the World."

There is considerable fresh evidence to the contrary, notably the revival of Islam and its emergence as a force working against the annexation of societies such as Iran by the Western way of life. The continuing strength of separatist movements, like those of the Basques and the Kurds, show how even a well-established political and military order can be shaken by new frontiers being erected inside a nation-state. A realistic appraisal of these trends indicates that the global village stereotype is merely a form of ideological manipulation meant to obscure the presence in the world of serious conflicts of interest—not to speak of the historical struggle between antagonistic classes and two social systems, capitalism and socialism.

REAFFIRMING PRINCIPLES

However attractive it may be for a Western eye to perceive the world of the twenty-first century in terms of a universal technoculture, it is wishful thinking rather than a serious prognosis. Despite communications, the world's divisions continue, not only between nation-states but often within societies.

Adequate preparation for the future cannot be based on a vision of Western life projected into the global village as a homogeneous marketplace for free enterprise—a sort of eighteenth-century dream of Adam Smith to come true in the twenty-first century. If realism suggests that we think in terms of individual societies, their historically determined formation, and their sociocultural peculiarities, then we should prepare ourselves for international cooperation in a democratic community of sovereign societies. This means more international equality and more mutual benefit instead of dependency and domination in the future family of nations.

Let us recall how the majority of today's nations see the global objectives. Here is a key statement from the final declaration of the 1979 summit meeting of nonaligned countries in Havana:

> The Sixth Conference of Heads of State or Government appeals to all peoples of the world to participate in efforts to free the world from war, the policy of force, blocs and bloc politics, military bases, pacts and interlocking alliances, the policy of domination and hegemony, inequalities and oppression, injustice and poverty and to create a new order based on peaceful coexistence, mutual cooperation and friendship, an order in which each people may determine its own future, attain its political sovereignty and promote its own free economic and social development, without interference, pressures or threats of any kind.[3]

This means no more and no less than a reaffirmation of the ideals codified in the charter of the United Nations and in other principles laid down for healthy international relations. In other words, there are good grounds to believe that international law and order, formulated in the twentieth century, will finally be put into practice in the twenty-first century.

Although this presents no revolutionary view of the future, it does imply painful surgery in a number of sensitive areas of the western hemisphere. The future course of communications will be determined by how the objectives for a new international information order are set forth by the movement of the nonaligned countries. These objectives are based on the following tenets:

(a) The fundamental principles of international law, notably self-determination of peoples, sovereign equality of states, and noninterference in the affairs of other states.

(b) The right of every nation to develop its own independent information system and to protect its national sovereignty and cultural identity, in particular by regulating the activities of transnational corporations.

(c) The right of people and individuals to acquire an objective [view] of reality by means of accurate and comprehensive information as well as to express themselves freely through various media of culture and communication.

(d) The right of every nation to use its information to make known its interests, aspirations, and political, moral, and cultural values.

(e) The right of every nation to participate on governmental and nongovernmental levels in the international exchange of information under favorable conditions, which provide a sense of equality, justice, and mutual advantage.

(f) The responsibility of various [participants] in the process of information to help [achieve] its truthfulness and objectivity as well as the particular social objectives to which the information activities are dedicated.[4]

These principles are endorsed by the United Nations, but they are still considered in the West to be far too radical. This was evident in the debate of the MacBride Commission Report at the General Conference of UNESCO in Belgrade in the autumn of 1980. The nonaligned proposal which contained the principles outlined above was forcefully resisted by the Western Bloc, led by the United Kingdom and the United States, with the result that the final compromise resolution was heavily watered down and flavored with "free flow" phraseology. It is to be noted, however, that the consensus reached at UNESCO does not compromise what obviously is most essential in the objectives of the new order: that it should be based on the fundamental principles of international law. I have examined the relation of these principles to journalism elsewhere.[5]

Thus the cornerstone of what is understood to be a new international information order (in short, NIIO) is the principle of national sovereignty as stated in point (a) of the above resolution. It has often been argued, especially by spokespersons for the U.S. press, that this means legitimizing authoritarian information policies, government control, and censorship of the press. Such accusations are based on ignorance of Third World objectives and the fundamentals of international law or on deliberate manipulation of public opinion. By no means is the collaboration and establishment of NIIO intended to undermine freedom of information. On the contrary, it is a program to achieve true freedom and pluralism with all nations and shades of opinion having reasonable representation in the total flow of information circulating in the world, a flow that today is characterized by glaring imbalances. Moreover, an increasingly vital component of NIIO has been recognized as the democratization of communication, not only at the international but, above all, at the national level. To put it bluntly it is an untruth to claim that NIIO will simply replace the present international system of dominance and dependency with another system in which nations are deprived of democracy. On the other hand, it *is* true that the concepts of democracy and freedom involved in NIIO—as well as in many other projects promoted by the UN—do not always coincide with the particular private-enterprise version of democracy and freedom so dear to Americans.

ON THE DECK OF THE *TITANIC*

If it is naïve to believe that tomorrow will bring us a communications age in the global village, it is equally naïve to believe that a versatile and egalitarian community of nations with democratic rules of international behavior will emerge without a bitter struggle. Tomorrow will bring us difficult times that will be at least as complicated as any experienced in the late twentieth century.

Times are getting worse and will continue to do so unless a key problem facing humankind is solved: How to put an end to preparation for a nuclear holocaust and the use of military force to solve international conflicts. The arms race, especially in strategic nuclear weapons, constitutes a global problem of such dimensions that it supersedes all other questions of life and death facing humankind today. Take the hunger and disease that face two-thirds of the world's people: Is this not a paradox in a world in which all development and emergency aid represents only 5 percent of the combined expenditures for armament? The stark fact is that we—or rather the militaristic forces dominating policies in our countries—are 95 percent more interested in destroying life on this planet than in safeguarding the preconditions of human survival. Indeed the North-South problem actually proves to contain much of the East-West problem. There is a growing realization that no real progress toward a solution of the global problems can be made unless a determined end is put to the arms race and disarmament is gradually achieved. That this has received recognition in contemporary Western thinking is outstandingly indicated by the report of the Brandt Commission.[6]

One might ask what such an alarming global perspective has to do with communications in the twenty-first century. A general answer to this question can be extracted from what has just been said. Communications cannot be separated from the rest of the socioeconomic-political issues: the more crucial these issues, the greater their relevance to the field of communications. But there are further reasons why the arms race and disarmament are especially pertinent to the considerations of this book.

First, in the West the mechanism of the arms race includes the mobilization of public opinion in support of increased defense spending. The mass media are crucial instruments in this process of manipulation. Developments in the United States provide a cardinal example of how the "free" media are becoming increasingly controlled by the militaristic orientation of the government. I do not mean to suggest that U.S. journalists are deliberately serving as agents of the Pentagon. They are almost invariably honest people devoted to their profession and with no strings attached. Nor are most of the media led around by the Pentagon. The relationship is a more subtle one. It is based on shared values and on uncritically (even unconsciously) accepted practices that lead to a symbiotic relationship with the government—or rather with the military–industrial complex behind the Pentagon.

The selling of the Pentagon, as the phrase has it, has proceeded effectively with the acquiescence of the media. I cite a recent editorial in the *Columbia Journalism Review* that criticizes the handling by the press of Presidential Directive 59 on limited nuclear warfare issued by the Carter Administration. The editors endorse the view of Fred Kaplan, author of the lead article in that issue, that "in the process of becoming explicators of Pentagon strategy, [reporters] seem to forget about the 'real world of nuclear warfare—its messy uncertainties, things going wrong, tens of millions of people dying, whole societies obliterated.' "[7]

What I ask is a review of the historically significant defense spending that took place in the United States and in the North American Treaty Organization countries. Where was the press as the fourth branch of government? Where was the adversary role of the press? Where was the civil disobedience that Elie Abel so elegantly describes?

Second, the industries that produce and maintain the communications infrastructures of the electronic age have merged with what former President Eisenhower called the military–industrial complex. Thus the contents of the mass media are conditioned to start with military interests. Further, the technology—and in many cases also the management of the communications industries—must depend increasingly on what is good for "defense." In other words, to unmask the real purpose, for destroying life and killing people.

Consequently, to speak about the neutron bomb or the SALT treaty in the context of communications in the twenty-first century is not only relevant but essential. We must be concerned not only morally but intellectually, and we must be seriously concerned because the situation is alarming. The *Titanic* of today's international community is sailing in dangerous waters; there may not be much time left to alter the course away from confrontation and toward peaceful cooperation and global solidarity. To do otherwise amounts to playing electronic games on the deck of the *Titanic*.

If such a viewpoint appears to be even remotely "political," this only proves how pervasive and dangerous is the dominant thinking of the West and how urgent is the task of adopting a more critical and realistic perspective.

Notes

1. Douglas Cater, "The Survival of Human Values," *The Journal of Communication* (Winter 1981).

2. François Régis Hutin, *Inter-Media* (January 1981).

3. Sixth Conference of Heads of State or Government of Nonaligned Countries, Havana, September 3–9, 1979; Final declaration, political part, paragraph 9 published, e.g., in *Review of International Affairs*, Belgrade, **30**, (707) 19 (September 20, 1979).

4. Fourth Meeting of the Intergovernmental Coordinating Council for Information of the Nonaligned Countries, Baghdad, June 5–7, 1980; Special Resolution on the New International Information Order, from Part I published, e.g., in *Communication in the Eighties, A Reader on the "MacBride Report,"* Cees Hamelink, Ed., IDOC International, Rome, 1980, pp. 31–32.

5. Kaarle Nordenstreng, *The Mass Media Declaration of UNESCO*, Norwood, Aglex, 1981.

6. Willy Brandt and Anthony Sampson, Eds., *North-South: A Program for Survival* (*The Brandt Report*), MIT Press, Cambridge, 1980.

7. "The Press's Own War Games," *Columbia Journalism Review*, 19 (January–February 1981).

INTELSAT V, the international communications satellite whose orbital sequence is shown above, can transmit 12,000 simultaneous telephone calls.

EPILOGUE

LOOKING BACK FROM THE TWENTY-FIRST CENTURY

ASA BRIGGS

It is a daunting assignment to have the last word. The agenda for this book reveals a sense of imaginative organization. Elie Abel looked forward from the twentieth century. I am given the task of pretending that I am living in the twenty-first century and looking back at our troubled time. It is a little like being asked to be a double agent for posterity and the present.

If the book was conceived in a spirit of imagination and symmetry, my task in conclusion is to offer perspective and balance; in this field both tend to be in short supply. So much in the communications world is instant, including instant fame. A week is a long time in politics, and a decade can acquire more of an identity in our time than a century used to do. We green America, and then we quickly gray it. As for balance, communications is a field in which most of the memorable generalizations are skewed, from McLuhan's "medium is the message" onward. And we know that some of the content and many of the effects are skewed, too. As John Robinson showed us, this includes the effects of television on the allocation of time in our personal time budgets.

Of one thing I am more sure than ever, it is impossible to discuss communications adequately. The same technology is organized and used in different ways in different societies and cultures, even in different offices. To understand why this is so we must make comparative studies not only of technologies and institutional structures but also of different social and cultural traditions. We are confronted with big differences between adjacent countries such as Britain and France, not to mention the even bigger differences between, say, Japan and the United States. We cannot therefore treat history as so much baggage to be discarded. It is a factor in the analysis. In many parts of the world what is most crucial is long-term history, not recent history. We are not merely dealing with the relationship between two centuries but among all the centuries.

This closing chapter therefore concentrates on establishing perspective. Elie Abel at the beginning and I at the end are both seeking it from the vantage point of the present, an uncertain position indeed. For whatever else our age may be called, the only thing certain is that it is a time of uncertainty. From this we must extract that sense of perspective which is a preoccupation shared by the historian and the futurologist.

NEW WAYS OF LOOKING AT OUR CENTURY

Where, then, do we stand at the present? To this question there can be no simple answer, for there is no one definitive version of the past any more than there is one definitive version of the future. The shapes of the past and the future change not only according to the amount of information available to us but according to our vantage points in time and space—whether we come from Finland or Minnesota. They vary also with the mood of the individual historian or futurologist, not to mention his or her concerns and values. There is at least as big a difference between Arnold Toynbee and Marc Bloch, the great French historian, who is one of my favorites, as there is between Peter Drucker and Alvin Toffler. We continually repaint the scene, both past and future, as we gain new insight into the present.

At some point in the future, though not early in the twenty-first century, it will be possible to play back periods of time stored in verbal and visual data banks to recapture immediacy and the feel of what it was like to live then. This will be made possible by highly sophisticated communications technology. We already do this in limited ways with films and music, including pop. We do it with fiction, and we do it by collecting objects, including those vintage treasures that have become so greatly sought after in the last years of the twentieth century. But we have not yet achieved what I think of as total immersion in the past; this may not come until the end of the next century. When we do achieve it, our present century will seem like an initial breakthrough period when, for the first time, the technology of recording words and pictures, notably pictures, became a reality. These technical attainments, when combined with mathematics, created new modes of communication that made it possible to gain access to the past just as they opened access to space.

We will therefore have the opportunity in the twenty-first century of becoming explorers in both space and time. It may be that the sheer bulk of preserved archives in the twenty-first century will be so overwhelming that except for diversion, people will be discouraged from looking too closely into the past Not that I underrate the propensity to be diverted or entertained. It certainly has been a major force in this century, which is reflected in the development of film and television and also in the enormous popularity of spectator sports. I believe that in the twenty-first century there will be more games rather than fewer of them, a remark that will warn the reader that I have not yet fully taken over my double-agent role; I am still looking at the twenty-first century as the future.

It is likely that there will be some people living in the next century who will reject what one author in this book referred to as love affairs with the future. In the same way Brian Aldiss, the British science-fiction novelist, identifies the craving to recapture the past as an affliction that strikes down so many of us before we reach middle age. In our century there have been some people who looked back to the nineteenth century as though they were displaced persons in their own time. They will undoubtedly have counterparts in the twenty-first—people who suffer from the twenty-first-century version of future shock. There almost certainly will be a nostalgic search for vintage objects and symbols, perhaps from our own period, which by then may be seen as an era of enormous challenge, of raw effort both mental and physical, and of predominantly human rather than machine concerns.

Recovering immediacy is one aspect of the approach to the past. We also have a desire when we look at the past to put it securely in place, to provide a framework for it. Just as futurologists cannot look ahead adequately by projection or extrapolation (certainly not with those old curves pointing to heaven or hell), so in the twenty-first century people will not be able to look back adequately on the twentieth century simply by tracing lines of cause and effect. They will have to look for patterns and for what has come to be known in the late twentieth century as "networks." (I had never heard this key concept used until about twenty years ago.) For continuities and discontinuities, and for contradictions, historians naturally will still be interested in the details of particular events, including Elihu Katz's media events. But they will want to relate these events to processes and periods.

We can trace some kind of periodization of the twentieth century even before we get into the twenty-first. The twentieth began with an era sometimes called the "belle époque," although there is some question whether for most of the people living before 1914 it can truly be called that. It was followed by the period of wars, with the two world wars so close to each other that future historians may tend superficially to treat them together. There can be said to have been a second belle époque, stretching from the late 1940s to the late 1960s and perhaps into the early 1970s. In the 1960s it was fashionable to think that a great change, a kind of biological-socio-logical-historical mutation would follow in the late twentieth century. I now feel that Herman Kahn's phrase, "époque de malaise," may be a more fitting name for the age in which we live. In each of the different periods of the century we have looked differently at the past and the future.

As several observers have pointed out, we always got certain crucial things wrong. In particular, economists of the second belle époque did not foresee the coincidence of inflation and mass unemployment that would grip many parts of the world today. Sociologists did not foresee the revival of religion, a development that makes Anne Branscomb's comment that we can choose between varieties of religion and lifestyle somewhat parochial. Nor did political scientists foresee the vehemence of nationalism, particularly minority nationalism. Looking back, we can now conclude that the second belle époque was not a very good time for prediction. This suggests that in the twenty-first century there doubtless will be many different and precarious vantage points just as there have been in this century.

During the first two or three decades of the new century the people making some of the key decisions will be those who have already finished their formal education. These people therefore will belong to this century as well as the next; they will be the bridge generation. In that sense the future is already here, as it is in other ways. Some trends, notably demographic, seem already to have been set. Certain managerial and political decisions concerning the twenty-first century, given the time lag in implementation, have already been taken. New generations will follow, however, that never knew what it was like to live in the twentieth century; they will be looking forward with mixed emotions to the twenty-second century.

EIGHT FEATURES OF THE TWENTIETH CENTURY

At this point I slip into my role as a double agent looking back at the twentieth century. As the twenty-first century progresses, the pictures of our time will be changing constantly. I can, however, identify eight salient features of the twentieth century that I feel will still stand up well in years to come. Although only one of the eight is solely concerned with communications, all have a communications dimension and an educational aspect as well.

When people look back at this century, the first thing that will impress them will be the fantastic geopolitical shifts that have occurred. They will note that the significance of Europe, so great at the beginning of the century, diminished relatively and even absolutely; that the United States meanwhile rose to a position unthought of at the start of the century; and that Russia rose to a position that was hardly conceivable at the end of World War I. At the same time it will be seen that Japan proved the possibilities of industrializing effectively, even innovatively, as its performance in the last part of the century has demonstrated.

The twentieth century has seen the birth of international institutions, both governmental and private. The first serious effort at intergovernmental organization, the League of Nations, turned out to be a casualty, but multinational corporations are very much alive. Economic shifts associated with this development across the Atlantic among the industrialized countries have had important consequences for the developing world. From 1951 to 1972 the terms of trade improved 24 percent for the advanced countries. These terms were reversed in the case of wheat in 1972 and, dramatically, in the case of oil in 1973. We have been conscious since then of living in an epoch of malaise. Yet there remains a division of interest and outlook between North and South that preoccupies many people in our time.

There is an important communications dimension to these geopolitical shifts, and the question is, how will this appear to the people of the next century? It seems to me that it will be concluded that we lacked an adequate international infrastructure.

As we all know, communications are incipiently global. The technologies are all the same, yet the gulfs between communication zones are wide. Some parts of the world are actually communication dependencies in the sense that the content of their communications systems is not independently determined. It is becoming increasingly important to be able to chart the flow of world communications; this has been one of my preoccupations at the International Institute of Communications, and I feel that it is essential to get the picture right.

I agree with those who say that language is not the most important thing that divides us. Many other obstacles prevent us from building a more effective international infrastructure for exploiting communications than we have at present. We must create an infrastructure based on shared concern, not on opinion. Success in this effort will depend on our capacity to discuss issues intelligently and responsively across boundaries, not relying on slogans.

The second feature that people will see on looking back on the twentieth century relates to the foregoing discussion: This has been a century of war. Two world wars were followed by a peace that has seen huge military expenditures on both sides, hardly a peace that most people across the frontiers desire. We are aware of the continuing difficulties of communicating on issues related to these subjects within and between countries. The problems are personal as well as electronic. We handle them awkwardly, and almost every day we make what seem to me to be terrible and sometimes terrifying mistakes; the twenty-first century will know the outcome of our efforts.

The third outstanding feature of the twentieth century has been the great increase in material wealth. Granted this wealth is unequally apportioned between one group and another and one part of the world and another. Nevertheless, it has reduced the endemic poverty that characterized preindustrial and early industrial societies in many parts of the world. A full understanding of the implications of this great increase awaits the analysis that will eventually be made by historians of various highly complex twentieth-century phenomena that we only imperfectly comprehend. Among them are inflation, mass unemployment, unionization, the relationship between public and private sectors in different economics, the interaction of government and enterprise, and patterns of corporate management.

The role of communications in this third cluster also awaits full analysis. People from all social groups have been able to own, hire, or share the use of the new communications devices of the twentieth century: devices that have greatly influenced expectations and aspirations as well as the behavioral patterns of everyday life. The historian of the future will know far more than we can possibly know about the effects of the new computer-based technologies on patterns of work and leisure. What is transpiring here is still obscure, at least to me. It seems certain, however, that the way in which the communications dimension of leisure is related to other aspects of leisure (and work) will still be much discussed in the new century.

The fourth feature of our own century that will stand out when people look back is demographic change. The huge growth in world population has been modified by varying rates of growth as the century has gone on. For the first time in human history, ways were developed for controlling population growth that did not involve the old Malthusian restraints. The outcome of this development will become clear only in the twenty-first century; one need hardly point out the various possibilities.

The fifth distinguishing feature is the growth of knowledge in which quantum leaps occurred in physics earlier in this century and in the life sciences later on. The latter development, I believe, will be a particular preoccupation of the people living in the next century because of the choices that they will have to make—very difficult choices about bioengineering. Indeed these biological issues will be so important in the twenty-first century that there will be a looking back to many of the problems we are facing now in this rather fumbling period; it may look very old-fashioned.

The sixth feature of this century that will impress people in the future is the increase in institutional education. But will they be pleased with what we did with it?

Beginning with the elementary and secondary schools, we improved access and opportunity at enormously increased cost, and in the 1960s this continued with the dramatic doubling or more of higher education. It was only in the last quarter of the century that emphasis was finally given to life-long education. I agree with Isaac Asimov that it will prove to be the most important development of all—that is why I am chancellor of the Open

University in Britain. In my view the failure to develop life-long education earlier and to reallocate national resources accordingly will stand out as a failure of the twentieth century.

I believe that strategic education inside business is also tremendously important. Not only will this eliminate some elements of conflict and help individuals to adapt but it will also draw people into a much closer understanding of what is happening in their world.

The seventh important development—note that I put it only seventh—will turn out to be the compunications revolution. This is a basic technological revolution, comparable to the Industrial Revolution and to earlier eras of radical change in human history that involved the mind as well as the body. This newest revolution has had but little influence as yet on institutionalized education, but the interaction is now beginning to speed up more dramatically. The communications revolution up to this point has been geared a great deal to entertainment and news. I believe that people in the next century will see this trend as only one phase.

We hear a great deal about passing through new phases, each posing new connections, new opportunities, and new problems. But I am not sure that in my twenty-first-century persona I am going to see some of the predicted effects. I am skeptical about the sociopsychological soundness of the forecasts of increased individual consumer choice. I would have to know a great deal more about the kinds of attitude that will be brought to bear by individuals in making these choices. I am critical of some of the forms of concentrated ownership in this area and particularly of the range of content of software. Concern has been expressed about what has happened to literacy in this century. We have good reason, I think, to worry about visual as well as verbal literacy. In a world of pictures we have not learned the simple ABCs of interpretation and criticism.

Elihu Katz makes an important distinction when he compares individuating or individualizing aspects of the media with their integrating aspects. But it is not always a sharp one. I am convinced that in these extremes we will develop intermediate package media for target groups. I envision just as much packaging in this area as now exists, for example, in tourism and in many other features of our social and cultural lives. Yet I am not afraid that in the twenty-first century we will drop shared broadcasting, at least not of live events. Nor am I afraid that we are going to produce a homogenized *homo universalis* in that century who is really going to be regarded as having broken with all the legacies of the past.

The eighth feature that will command attention is twentieth-century culture. It is fairly easy for a scholar to discuss the culture of the Renaissance or even the culture of the Victorian Age. But what a culture ours is to describe! It has been, as I remarked earlier, a great age for the performing arts. There have been great novelists, film makers, painters, and, at least for part of the century, musicians. Yet in many ways there has been more stratification rather than less, with artists in the middle of the muddle, at best expressing rather than communicating. We would all wish, as Kathleen Nolan wishes, to get the position of the creative artist right in our society and our culture. But I doubt whether the twenty-first century will feel that we succeeded in doing so, at least not in the communications world. So far our most interesting successes appear to lie at the community level.

All eight features will be seen in the twenty-first century as having been interrelated. This perspective will make it impossible to miss the underlying economic factors; technology will never be the only determinant. There will also be a sense of involvement in continuing processes that stretch back to the nineteenth century and are still active in the twenty-first. Some of the changes that transpired in the twentieth century will seem even more rapid, I believe, than those occurring in the twenty-first century. Certain dramatic patterns of change that have occurred in this century will not necessarily be repeated in the next.

Although much of the geopolitical change that has taken place has been intriguing, I do not believe, for instance, that the map of Africa as it now exists is likely to survive into the twenty-first century. Nor do I believe any "new order" will easily establish itself. There will be conflict between establishments and minorities, and there is scope for plenty of surprises. I would not like a world without surprises.

The issues are complex and controversial, and in abandoning my double-agent role I should like to emphasize that we can face the future effectively—or for that matter comprehend the past—only if we bring people from different disciplines together with those who are involved directly in production and distribution. We will best be able to face the twenty-first century if we recognize that the future of the societies we represent, societies that prize freedom, depends in the short run on the mass of scattered moves that we make as individuals, both separately and in organizations.

Elie Abel says that the next ten or fifteen years could be decisive. If this is true, and I agree that it is, we will need more opportunities to get together to identify and assess objectives and priorities. It was many years ago that

I first heard the phrase "toward the year 2000" from Daniel Bell. By now we have been thinking about it so long that we may forget that we are separated from the year 2000 by less than the space of one generation. In these few years we need to ponder not only the communications revolution but the larger agenda that must be shaped for the world. The media must not let us reach the year 2000 without briefing us well about both facts and options. That is their responsibility.

APPENDIX

PATTERNS OF CHANGE IN THE INFORMATION AGE

Chart 1. The Sequence of Inventions in Telecommunications, 1840–2000

Although intercity telecommunications systems have had a tenfold capacity increase every twenty years since the introduction of the electric telegraph, the average capital cost of building these systems has dropped ten times in the last fifty years.

Chart 2. Computer User's Development Costs

Until recently most of the cost for manufacturing computer systems went for hardware research and design. But microelectronic technology has so reduced the expense of systems components and programming has become so much more complex that in the future more than 90% of computer costs will be for developing the software.

Chart 3. Worldwide Growth of Online Computer Storage

In the last five years the increase in data which can be accessed by computers without any humans loading magnetic tapes or discs has been almost one hundred-fold. It is predicted that by 1983 about one quadrillion bytes of disc memory—one byte equal to about one text character—will be connected to computers simultaneously.

Chart 4. Changing Work Patterns of Households

In 1960 the "conventional" family—a working male head-of-household and wife at home—accounted for 43 percent of all families. By 1990 only 14% of households are expected to resemble this nuclear pattern. By then more than three-quarters of all working families will have two members employed and 45 percent of all households will have a single head.

Chart 5.   Annual Factory Shipments: Communications and Electronic Equipment by Type, U.S., 1970–1979

Electronics for information processing is the largest segment of the electronics industry, but the demand for communications common carriers has been growing the fastest in the last half-decade. That growth may be accelerated even more by future demands from home information systems.

Chart 6.   U.S. Consumer Spending on Mass Communications, 1968–1978

Despite the steady growth of print, the electronic media—radio, TV, recordings—continue to represent the dominant proportion of consumer spending for entertainment and educational media.

Chart 7.   Trends in Personal Consumption Expenditures, 1960–1990

For decades expenditures for telecommunications followed the general growth pattern of all personal consumption, but in the last twenty years they have been rising at a faster rate. In contrast, personal transport has exhibited a traditional growth rate, but, if rising fuel and vehicle costs cause people to travel less, transportation expenditures could well level off.

Chart 8.   Type of Employment by Industry Sector

By the year 2000 service-producing industries will account for 70 percent of all jobs. At the same time computer-aided production will play an ever-increasing role in manufacturing, agriculture, and mining.

Chart 9.   Percent Penetration in U.S. Homes of Selected Consumer Communications Devices

The growth of the telephone and over-the-air telecommunications services had virtually saturated the U.S. market by 1960. Cable TV, with most systems having a two-way capacity and offering data services, is expected to be the fastest growing communications technology in our urban areas by the end of the twentieth century.

Chart 1

The Sequence of Inventions in Telecommunications, 1840-2000

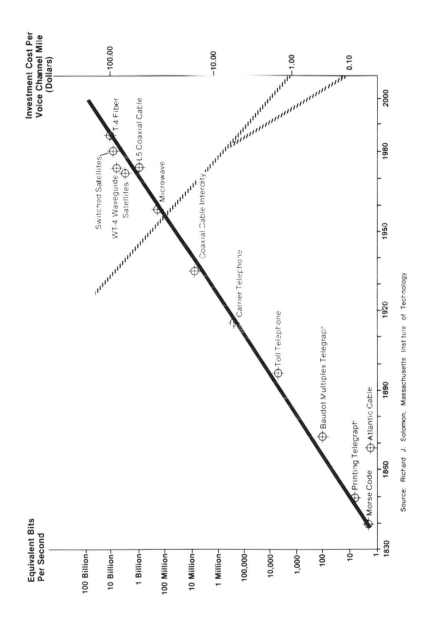

Source: Richard J. Solomon, Massachusetts Institute of Technology

213

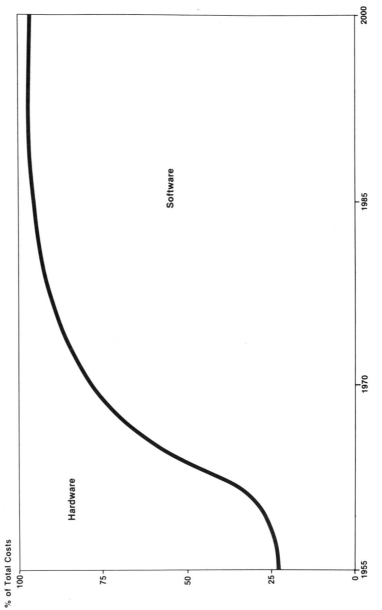

Chart 2

Computer User's Development Costs

% of Total Costs

Hardware

Software

Source: **Worldwide Semiconductor Industry, 1977**

214

Chart 3

Worldwide Growth of Online Computer Storage

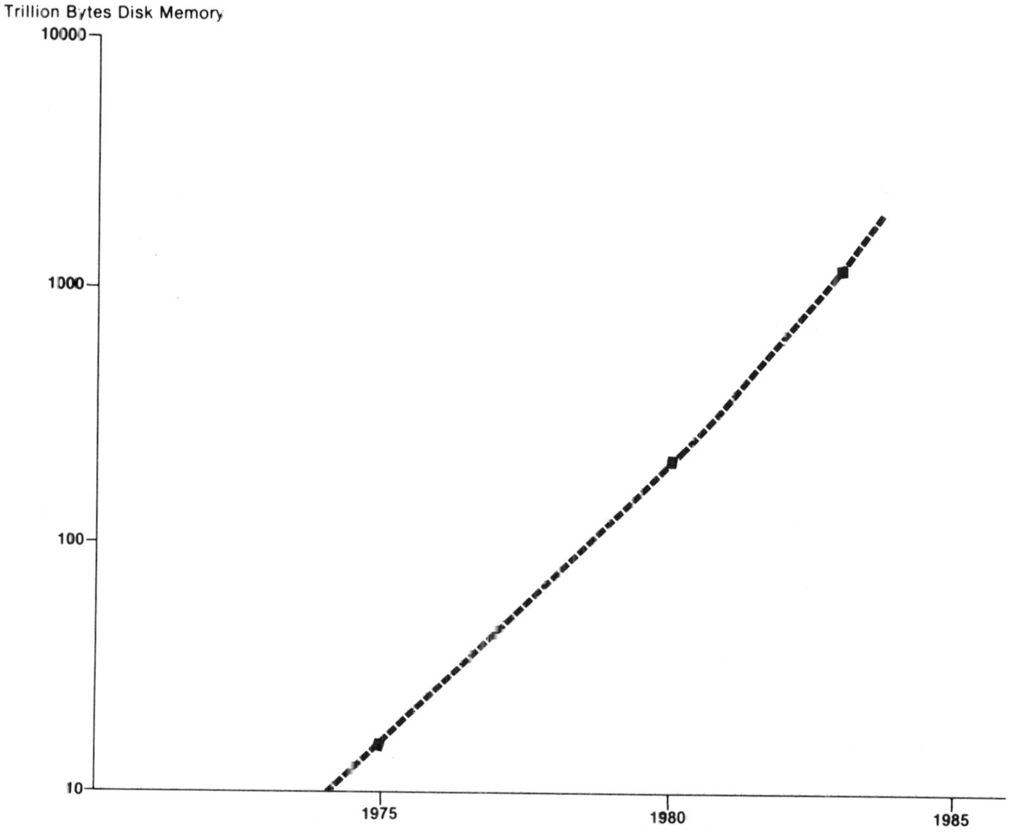

Trillion Bytes Disk Memory

10000

1000

100

10

1975   1980   1985

Source: IBM Corp.

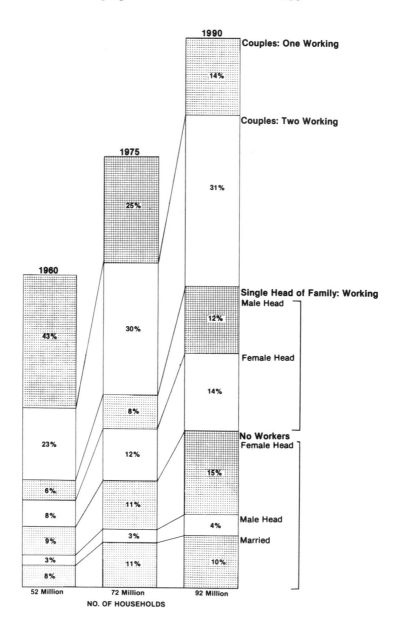

Chart 4   **Changing Work Patterns of Households**

1990
Couples: One Working

14%

Couples: Two Working

31%

Single Head of Family: Working
Male Head

12%

Female Head

14%

No Workers
Female Head

15%

Male Head

4%

Married

10%

1975

25%

30%

8%

12%

11%

3%

11%

1960

43%

23%

6%

8%

9%

3%

8%

52 Million   72 Million   92 Million

NO. OF HOUSEHOLDS

Source: Harvard—M.I.T. Joint Center for Urban Studies

216

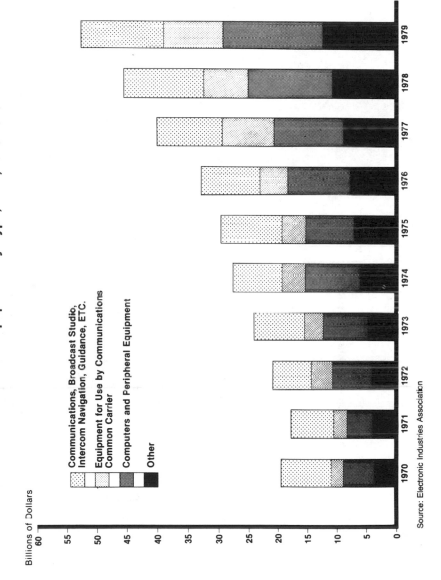

Chart 5

Annual Factory Shipments:Communications and Electronic Equipment by Type, U.S., 1970-1979

Billions of Dollars

Communications, Broadcast Studio,
Intercom Navigation, Guidance, ETC.

Equipment for Use by Communications
Common Carrier

Computers and Peripheral Equipment

Other

Source: Electronic Industries Association

217

Chart 6   **U.S. Consumer Spending on Mass Communications, 1968-1978**

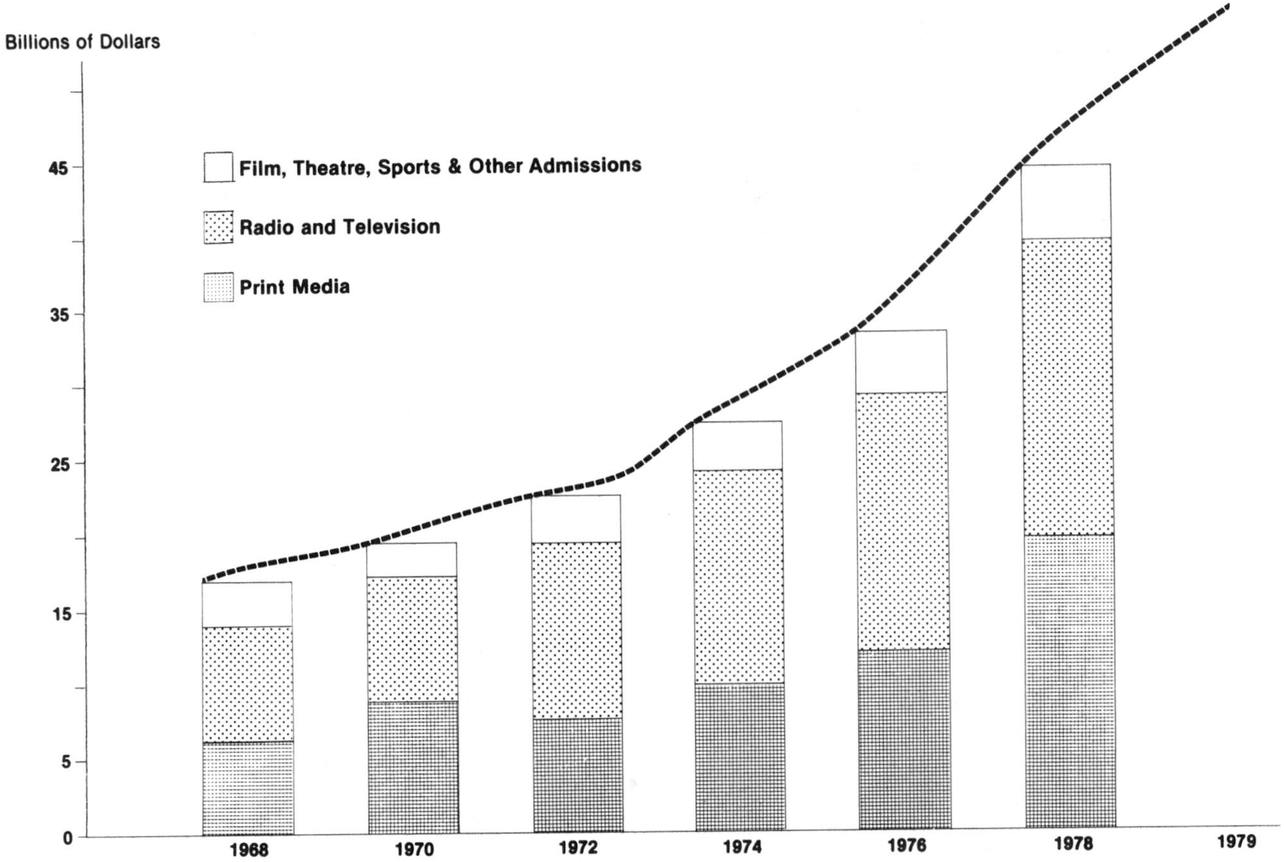

Billions of Dollars

Film, Theatre, Sports & Other Admissions

Radio and Television

Print Media

45

35

25

15

5

0

1968   1970   1972   1974   1976   1978   1979

Source: U.S. Department of Commerce, The National Income
and Product Accounts

Trends in Personal Consumption Expenditures, 1960-1990
(Constant Dollars)

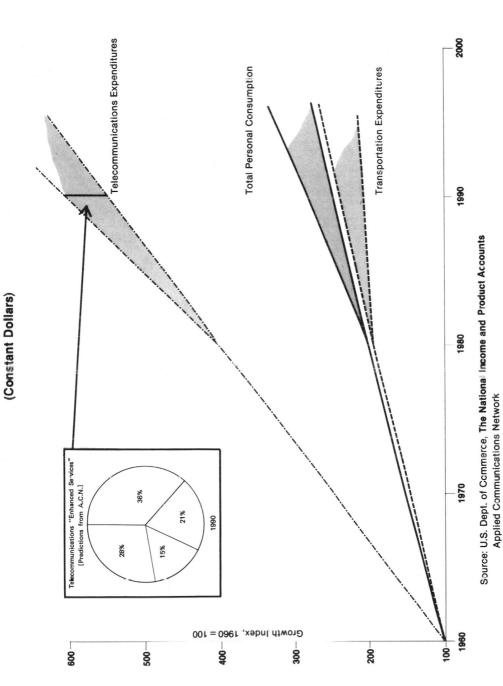

Telecommunications Expenditures

Total Personal Consumption

Transportation Expenditures

Growth Index, 1960 = 100

600 — 500 — 400 — 300 — 200 — 100

1960 1970 1980 1990 2000

Telecommunications "Enhanced Services"
[Predictions from A.C.N.]

36%
21%
15%
28%
1990

Source: U.S. Dept. of Commerce, The National Income and Product Accounts
Applied Communications Network

219

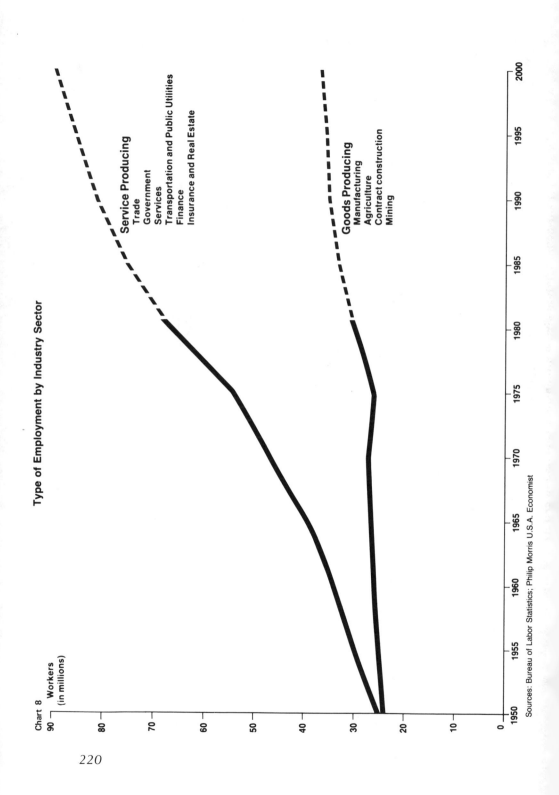

Chart 8

Type of Employment by Industry Sector

Workers
(in millions)

Service Producing
Trade
Government
Services
Transportation and Public Utilities
Finance
Insurance and Real Estate

Goods Producing
Manufacturing
Agriculture
Contract construction
Mining

90
80
70
60
50
40
30
20
10
0

1950 1955 1960 1965 1970 1975 1980 1985 1990 1995 2000

Sources: Bureau of Labor Statistics; Philip Morris U.S.A. Economist

220

Chart 9 **Percent Penetration in U.S Homes of Selected Consumer Communications Devices**

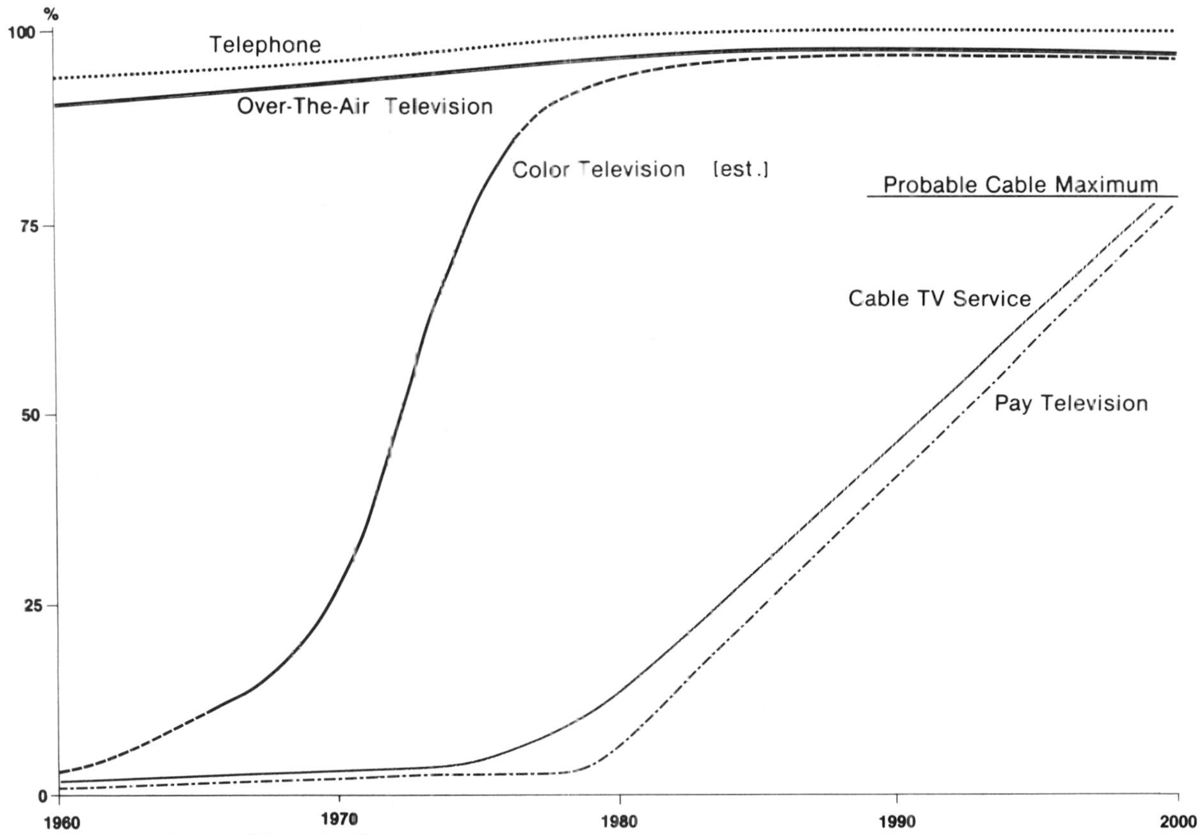
%
100
75
50
25
0
1960
1970
1980
1990
2000
Telephone
Over-The-Air Television
Color Television [est.]
Probable Cable Maximum
Cable TV Service
Pay Television

Sources: U.S. Dept. of Commerce
AT&T, "The World's Telephones"
Electronic Industries Association

BIBLIOGRAPHY

Bagdikian, Ben, *The Information Machines*, Harper & Row, New York, 1971.

Brandt, Willy and Anthony Sampson, Eds., *North-South: A Program for Survival* (*The Brandt Report*), MIT Press, Cambridge, 1980.

Cater, Douglas, "The Survival of Human Values," *The Journal of Communication* (Winter 1981).

Coen, Robert J., "Vast U.S. and Worldwide Ad Expenditures Expected," *Advertising Age*, **51** (49), 1–16 (November 13, 1980).

Confino, Michael, *Historical Consciousness in Contemporary Society*, The Aranne Foundation at the University of Tel Aviv, Tel Aviv, 1980.

Crawford, Morris H., "Toward an Information Age Debate," *Chronicle of International Communication*, **1** (2), 3 (December 1980).

DeMaio, Harry B., *Data Regulation: European and Third World Realities*, Online, Uxbridge, England, 1978.

Dertouzos, Michael L. and Joel Moss, Eds., *The Computer Age: A Twenty-Year View*, MIT Press, Cambridge, 1979.

Ginzberg, Eli and George J. Vojta, "The Service Sector of the U.S. Economy," *Scientific American*, **244** (3), 48–55 (March 1981).

International Information Flow: Forging A New Framework, Thirty-Second Report of the Committee on Government Operations, 96th Congress, 2nd Session, House Report No. 96–1535, December 11, 1980, U.S. Government Printing Office, Washington, D.C.

International Project for Soft Energy Paths, *Soft Energy Notes*, Friends of the Earth, San Francisco, 1978.

223

Jackson, Wes, *New Roots for Agriculture*, Friends of the Earth, San Francisco, 1980.

Katz, Elihu, "Media Events: The Sense of Occasion," *Studies in Visual Communication*, **6**, 84–89 (1980).

Katz, Elihu and Michael Gurevitch, *The Secularization of Leisure*, Faber, London, 1973.

Keen, Peter G. W., "Information Systems and Organizational Change," *Communications of the ACM*, **24** (1), 24–33 (January 1981).

Keen, Peter G. W. and M. S. Scott Morton, *Decision Support Systems*, Addison-Wesley, Reading, MA, 1978.

King, J. L. and K. L. Kraemer, "Costs as a Social Impact of Information Technology," in M. L. Moss, Ed., *Telecommunications and Productivity*, Addison-Wesley, Reading, MA, 1981.

Lovins, Amory B. and L. Hunter Lovins, *Energy Policies for Resilience and National Security*, report to President's Council on Environmental Quality for Federal Emergency Management Agency, 1981, in press.

Lovins, Amory B. and L. Hunter Lovins, *Energy/War: Breaking the Nuclear Link*, Friends of the Earth, San Francisco, 1980, and Harper & Row Colophon paperback, 1981.

Noguchi, Yukio, *Johoono Keizai Riron* (The Economic Theory of Information), Toyo Keizai Shinposha, Tokyo, 1977.

Nolan, R. L., "Managing the Crises in Data Processing," *Harvard Business Review*, **57**, 115–126 (March–April 1979).

Nordenstreng, Kaarle, *The Mass Media Declaration of UNESCO*, Aglex, Norwood, 1981.

Oettinger, Anthony G., "Information Resources: Knowledge and Power in the 21st Century," *Science*, **209**, 191–198 (July 4, 1980).

Pool, Ithiel de Sola, "Exporting Data—Latest Paranoia," *Telecommunications Policy*, **4**, 314 (December 1980).

Pool, Ithiel de Sola, Ed., *The Social Impact of the Telephone*, MIT Press, Cambridge, 1977.

Power, Philip H., "Threat to Ad Freedom?" *Advertising Age*, **51** (54), 44 (December 15, 1980).

"The Press's Own War Games," *Columbia Journalism Review*, 19 (January–February 1981).

Schiller, Herbert I., "Decolonization of Information: Efforts Toward a New International Order," *Latin American Perspectives*, (V), 1 (Winter 1978).

Sheppard, C. Stewart and Donald C. Carroll, Eds., *Working in the Twenty-First Century*, Wiley, New York, 1980.

Smith, Anthony, *The Geopolitics of Information*, Oxford, New York, 1980.

Stobaugh, Roger and Daniel Yergin, Eds., *Energy Future: The Report of the Harvard Business School Energy Project*, Random House, New York, 1979.

Tehranian, Majid, "Iran: Communication, Alienation, Revolution," *Intermedia*, **7**, 6–12 (1979).

Uno, Kimio, "The Communication Sector in Japan and Its Role in Economic Development," Workshop on the Economics of Communication, the East-West Communication Institute, June 1980.

Zuboff, S., "Psychological and Organizational Implications of Computer-Mediated Work," Center for Information Systems Research, Sloan School of Management, MIT, Working Paper No. 71, June 1981.

BIOGRAPHICAL NOTES

THE AUTHORS

ELIE ABEL, LL.D., Harry and Norman Chandler Professor of Communication, Stanford University. Dr. Abel has been active in communications for some forty years. His appointment to the Stanford faculty in 1979 was preceded by nine years as Godfrey Lowell Cabot Professor and Dean of the Graduate School of Journalism at Columbia University in New York.

Born in Canada in 1920, Dr. Abel began his career as a fledgling reporter for Canadian newspapers. After serving in the Royal Canadian Air Force in World War II, he was for several years a national and foreign correspondent for *The New York Times* and in 1959 he joined *The Detroit News* as chief of its Washington bureau. In 1961 he joined NBC's news division, first as State Department correspondent, then as London bureau chief, and finally as diplomatic correspondent. While at NBC, he won the George Foster Peabody Radio and Television Award and two Overseas Press Club awards.

Dr. Abel has authored or coauthored three books: *The Missile Crisis; Roots of Involvement: The U.S. in Asia* with Marvin Kalb; and *Special Envoy to Churchill and Stalin,* with W. Averell Harriman. He has served as a member of the MacBride Commission of UNESCO, which recently concluded a two-year study of international communications problems.

GAIL E. BERGSVEN, Vice President, Human Services Programs, Control Data Corporation. Ms. Bergsven has been with Control Data Corporation since 1969. She is currently an honorary board member of the National Committee for Prevention of Child Abuse, a member of the board of directors and secretary of the Stress

227

Resource Institute, and a member of the board of directors of the Girl Scout Council of St. Croix Valley. In recent years she has served as a member of the board of directors of the Twin Cities Personnel Association.

ANNE W. BRANSCOMB, J.D., communications consultant, New York. Mrs. Branscomb is an attorney specializing in communications law and presently serves as chairman of the Communications Committee of the Science and Technology Section of the American Bar Association. She has published numerous articles and reports on communications technology and public policies. She has also served on the U.S. Department of Commerce Technical Advisory Board and the U.S. Department of State WARC Advisory Committee. Mrs. Branscomb has been a trustee of Educom, a director of National Public Radio, communications counsel for the Teleprompter Corporation, and vice president and chairman of Kalba Bowen Associates, communications consultants.

ASA BRIGGS, Lord Briggs of Lewes, LL.D., Litt.D., D.Sc., Provost of Worcester College, Oxford, and Chancellor of the Open University. Lord Briggs, who became a life peer in 1976, is a historian specializing in nineteenth- and twentieth-century social and cultural history. His several books include *Victorian People, Victorian Cities, The Age of Improvement, The History of Broadcasting in the United Kingdom,* and *Governing the BBC.* He has been president of the British Social History Society since its founding in 1976 and in 1975 received the Marconi Medal for his work on communications history.

Lord Briggs has served on many national committees concerned with education and culture. He has been chairman of the Council of the Institute of Development Studies, vice chairman of the United Nations University, and a governor of the British Film Institute.

Lord Briggs is currently chairman of the Council of the European Institute of Education in Paris, chairman of the European Educational Research Trust in London, a trustee of Great Britain's Civic Trust, and chairman of the Leverhulme Trust Awards Committee.

HARRY L. FREEMAN, Senior Vice President, Office of the Chairman, American Express Company. Mr. Freeman's principal responsibility is worldwide corporate-level strategic development. He is also responsible for relations with international institutions and governments. Before joining American Express Mr. Freeman managed the project financing team worldwide of Bechtel Group Incorporated in San Francisco and was vice president of the Overseas Private Investment Corporation, a U.S. government agency. He has also practiced law in San Francisco, specializing in tax, corporate, and international areas.

LAWRENCE HALPRIN, landscape architect-planner; founder of Round House. Mr. Halprin's designs range from rapid transit systems to university campuses, from large-scale civic redevelopment to inner-city parks. He has written seven books and made two films, one of which is the award-winning documentary on Salvador Dali. His credits include several awards, most notably the American Institute of Architects' Gold Medal. Mr. Halprin is a fellow of the American Society of Landscape Architects and of the American Institute of Interior Designers. Since 1970 he has been an advisor of the National Endowment for the Arts, and since 1969 he has been a member of the Advisory Committee for the Planning of Jerusalem. He has also served on the National Council on the Arts and the Advisory Council on Historic Preservation.

AMORY B. LOVINS, D.Sc., physicist, and L. HUNTER LOVINS, J.D., sociologist and political scientist, are consultants in energy policy, working as a team in more than fifteen countries. They also serve as policy advisors to Friends of the Earth, Inc., a nonprofit conservation lobbying group.

In addition to his association with Friends of the Earth, Amory Lovins is vice president of FOE Foundation, a related educational charity organization. Dr. Lovins is also a member of the U.S. Energy Research Advisory Board. His books include *Soft Energy Paths* and *Energy/War: Breaking the Nuclear Link*.

From 1974 to 1979 L. Hunter Lovins was the assistant director of the California Conservation Project ("Tree People"), which she cofounded. In this project thousands of smog-tolerant trees were planted in forest and urban areas of southern California. She has lectured on and served as a consultant in urban forestry.

ELIHU KATZ, Ph.D., Professor of Sociology and Communication, The Communications Institute, Hebrew University of Jerusalem, and The Annenberg School of Communications, University of Southern California. Dr. Katz divides his time between The Communications Institute, of which he was founding director, and The Annenberg School. He has also served on the faculties of the University of Chicago and Columbia University. His most recent books include *The Secularization of Leisure*, with Michael Gurevitch; *Broadcasting in the Third World*, with E. G. Wedell; and *Almost Midnight: Reforming the Late-Night News*, with Itzhak Roeh, Akiba A. Cohen, and Barbie Zelizer. Dr. Katz was the founding director of Israel Television and was the first recipient of the German prize in international communications research, *In Medias Res*.

PETER G. W. KEEN, D.B.A., Associate Professor, Management Science, Sloan School of Management, Massachusetts Institute of Technology. In addition to teaching at The Wharton School, Stanford Graduate School of Business, Harvard Business School, and MIT, Dr. Keen is an editor, writer, and consultant. He is currently coeditor of the Addison-Wesley Publishing Co. Series on Decision Support Systems, a member of the editorial board of the *Human Systems Management* journal, and managing editor of *Office: People and Technology*. Dr. Keen has written and consulted on the implementation of computer systems, motivation, and management of organizational change.

LOUIS H. MERTES, Vice President and General Manager, Systems, Continental Bank. Mr. Mertes has been with Continental Bank for twelve years; he is currently responsible for a 750-person unit that directs and maintains the company's worldwide automation efforts. Before joining the bank he was a project manager for Motorola Inc., a sales representative for IBM Corporation, and an engineer for Western Electric Co., Inc. Mr. Mertes heads various civic and charitable organizations, which include the Illinois Society for the Prevention of Blindness.

J. RICHARD MUNRO, President and Chief Executive Officer, Time Inc. Mr. Munro joined Time Inc. in 1957 as a member of the circulation department staff of *Time* magazine. In 1960 he shifted to *Sports Illustrated* and was named publisher in 1969. Appointed a vice president in 1971, Mr. Munro had responsibilities for the company's book publishing, cable TV, and film operations. In 1975 he became group vice president for video and was named to his present position in October 1980. Mr. Munro is a member of The President's Council on Physical Fitness and Sport, a trustee of Experiment in International Living, and a director of the Urban League of Southwestern Fairfield County. He is also a trustee of Colgate University.

KATHLEEN NOLAN, actress; former President, Screen Actors Guild. Ms. Nolan has been active in the performing arts, both on stage and off, for most of her life. Her acting career ranges from an award-winning role as Wendy in *Peter Pan* on Broadway to the part of Kate in the long-running television series, *The Real McCoys*. Ms. Nolan served as president of the Screen Actors Guild from 1975 to 1979, the first woman to hold that office. She was also one of the first performers to tour Vietnam and received a Presidential Citation for that service. Ms. Nolan is currently a member of the board of directors of the Corporation for Public Broadcasting and is working on various theater and film projects as well as a book entitled *Woman as Artist and Activist*.

KAARLE NORDENSTRENG, Ph.D., Professor of Journalism and Mass Communication, University of Tampere, Finland. Born in Finland, Dr. Nordenstreng began writing professionally at age fifteen as a freelance journalist for Finnish radio; he has been deeply involved in communications ever since. Among other accomplishments, Dr. Nordenstreng has been a consultant to UNESCO on communications research and policies since 1969 and a vice president of the International Association for Mass Communication Research since 1972. He has served as a member of several national commissions on communications and foreign policy and has written or edited a dozen books, which include *National Sovereignty and International Communication: A Reader*, coedited with Herbert I. Schiller; and *Television Traffic—A One-Way Street?* coauthored with Tapio Varis.

ARNO A. PENZIAS, Ph.D., Nobel Laureate; Executive Director, Research, Communications Sciences Division, Bell Laboratories. Dr. Penzias joined Bell Laboratories in 1961, where he has conducted research in radio physics and astronomy, atmospheric physics, and most recently, the formation of simple chemical molecules in outer space. In 1978 he and Robert Wilson, also of Bell Labs, won the Nobel Prize for Physics for their 1964 discovery of evidence to support the "big bang" theory of the origin of the universe. Dr. Penzias has won numerous other honors, among which are the National Academy of Sciences' Henry Draper Medal and the Royal Astronomical Society's Herschel Medal (both in 1977). He is a member of several distinguished scientific societies, including the Astronomy Advisory Committee of the National Science Foundation. He is also a vice chairman of the Committee of Concerned Scientists, a national organization that works for the political freedom of scientists worldwide.

JOHN P. ROBINSON, Ph.D., Professor of Sociology; Director, Survey Research Center, University of Maryland. Dr. Robinson has written extensively on the role of mass communications in everyday life, including the interplay of television and leisure time, public comprehension of information conveyed in the media, the influence of mass media on voting behavior, and public interpretation of cultural media content. He has served as a study director of the University of Maryland's Institute for Social Research, director of the Communications Research Center at Cleveland State University, research coordinator for the U.S. Surgeon General's Committee on Television and Human Behavior, and research consultant to the BBC. Dr. Robinson is currently on the editorial boards of *Public Opinion Quarterly, Journal of Communication, Social Psychology Quarterly,* and *Media, Culture and Society.*

F. G. RODGERS, IBM Vice President, Marketing, International Business Machines Corporation. Mr. Rodgers, who has been with IBM for thirty-one years, divides his time between its marketing and data-processing activities. In 1967 he was named president of the Data Processing Division. In 1970 he was appointed IBM director of marketing, and in 1974 he assumed his present position. Mr. Rodgers is a trustee of the Marketing Science Institute at Harvard University, a Woodrow Wilson Visiting Fellow, and director and vice president of the Sales Executives Club of New York. He is also a member of the advisory councils of Miami University and the University of Tennessee.

HERBERT I. SCHILLER, Ph.D., Professor of Communications, Third College, University of California, San Diego. Dr. Schiller has been involved in economics and communications education for some forty years. He has taught in several American universities as well as in Jerusalem, Sweden, and Finland. His published work is voluminous and he is a contributing editor to several magazines in a number of countries. Dr. Schiller is a trustee of the International Institute of Communications in London, a vice president of the International Association for Mass Communication Research, and a member of the Academic Review Board, Latin American Institute for Transnational Studies. His latest book is entitled *Who Knows: Information in the Age of the Fortune 500.*

KIMIO UNO, Ph.D., Associate Professor, Institute of Socio-Economic Planning, University of Tsukuba, Japan. Dr. Uno, born in Tokyo, has been deeply involved in Japanese economic affairs as a member of The Council on Economic Welfare and the Bureau of Statistics and as an economist for The Foundation for the Advancement of International Science. Until recently he was the staff economist for the Japan Economic Research Center. Dr. Uno, who also spent a year as a visiting fellow at Yale University, is a member of several academic societies, has participated in numerous international conferences, and has written extensively on sociological and economic matters.

TIMOTHY E. WIRTH, U.S. Representative (D. Colorado). Congressman Wirth was elected to the U.S. House of Representatives in 1974. He is currently chairman of the Subcommittee on Telecommunications, Consumer Protection and Finance, which is updating the Communications Act of 1934. He is also a member of the Subcommittee on Fossil and Synthetic Fuels, the Democratic Steering and Policy Committee, and the House Committee on the Budget. In addition, Congressman Wirth organized the 94th Members' Caucus which has played a major role in House reform. Before becoming a congressman he served with the U.S. Department of Health, Education and Welfare in Washington. In 1970 he received that organization's Distinguished Service Award.

THE EDITORS

ROBERT W. HAIGH, D.C.S., Dean and Distinguished Professor of Business Administration, The Colgate Darden Graduate School of Business Administration, University of Virginia. Dr. Haigh is currently a member of the board of directors of Landmark Communications Inc. and The Institute of Chartered Financial Analysts. In past years he has been president of the Educational Information Publishing Group, group vice president and member of the board of directors of Xerox Corporation; vice president of The Standard Oil Company (Ohio) and president of three of its subsidiaries; and president or chief executive officer of several venture capital and consulting concerns. He has served as acting director of The Wharton Applied Research Center and taught for six years at Harvard Business School.

GEORGE GERBNER, Ph.D., Dean and Professor of Communications, The Annenberg School of Communications, University of Pennsylvania. Born in Hungary, Dr. Gerbner came to the United States in 1939. He was a member of the OSS in World War II and, before joining the University of Pennsylvania in 1964, taught at the University of Illinois and the University of Southern California, among other schools. The author of numerous articles, reports, and books on mass communications research, he is currently the editor of the *Journal of Communication*. Dr. Gerbner has directed multinational communications research projects for such organizations as the National Science Foundation, the U.S. Office of Education, and UNESCO. His current research on "Cultural Indicators," including the television "Violence Profile and Index," is conducted under a grant by the National Institute of Mental Health.

RICHARD B. BYRNE, Ph.D., Professor and former Acting Dean, The Annenberg School of Communications, University of Southern California. Dr. Byrne, who has been a photographer and television producer, has also designed exhibitions and written building programs for many small theaters and three communications buildings. He has served on several civic boards related to the arts and the special needs of the handicapped. Dr. Byrne is president of The Byrne Group, a Hollywood-based communications consulting firm.

INDEX